Hardy Cross

Hardy Cross: American Engineer

LEONARD K. EATON

University of Illinois Press
URBANA AND CHICAGO

© 2006 by the Board of Trustees
of the University of Illinois
All rights reserved

Supported by the Graham Foundation.

Manufactured in the United States of America
∞ This book is printed on acid-free paper.
C 5 4 3 2 1

Library of Congress Cataloging-in-Publication Data

Eaton, Leonard K.
Hardy Cross : American engineer / Leonard K. Eaton.
p.   cm.
Includes bibliographical references and index.
ISBN-13: 978-0-252-02989-9 (cloth : alk. paper)
ISBN-10: 0-252-02989-5 (cloth : alk. paper)
1. Cross, Hardy, 1885–1959. 2. Engineers—United States—Biography.
I. Title.
TA140.C77E28   2006
620.0092—dc22   2006015292

Contents

Preface   *vii*

1. The Preparatory Years, 1885–1921   *1*
2. The Creative Years at Illinois, 1921–37   *18*
3. The Years at Yale, 1937–51, and Retirement   *53*

Appendix A: Glossary   *89*

Appendix B: Selected and Annotated Bibliography of the Writings of Hardy Cross   *93*

Appendix C: Hardy Cross's Contribution to Structural Analysis   *101*
   Emory Kemp

Notes   *111*

Index   *115*

Preface

In the field of structural engineering, Hardy Cross was the outstanding American of his time (1885–1959). In the area of reinforced concrete his contribution was unsurpassed. His achievement was widely recognized in both the United States and Europe. This book is an attempt to give him back to his compatriots and fellow professionals.

The title of this book reflects my conviction that Hardy Cross was unique among American structural engineers of his day. The truly original engineers during Cross's lifetime were almost always European. Karl Terzaghi, the father of soil mechanics, came from the Italian portion of the former Austro-Hungarian Empire. He practiced his profession on both sides of the Atlantic. Theodore (Todor) von Karman, the most prominent aeronautical engineer of the period, was educated in Europe and took up a position as research associate at the California Institute of Technology when he was nearly fifty. Karl Westergaard, Cross's eminent contemporary at the University of Illinois, was of Danish origin and had worked in Germany. Eugen Freyssinet, the pioneer of reinforced concrete construction in France, m be viewed as a continuator of the grand tradition of the Ecole National Ponts et Chaussees. Stefan Timoshenko, author of numerous treati structural engineering and a man who had notable differences wit' was a Russian. This list may be extended substantially.

The point is made more forcefully if we consider the Manh ect, perhaps the most significant engineering task ever und United States. By 1939 the science behind the atomic bomb in Britain, the United States, and Germany. The British race to build the nuclear weapon until 1941, when the Ur its vast industrial resources into play. The theoretical physics section at Los Alamos was staffed mostly by men with German training. But the man who solved the engineering problem by bringing together the critical masses of U-235 was an Italian, Enrico Fermi. Hundreds of lesser engineers worked on the project, but Fermi's contribution was the key. The same general analysis could be made for the Radiation Laboratory at the Massachusetts Institute of Technology, but in that case the basic engineering solutions were devised by the British. In the company of truly creative twentieth-century engineers, Hardy Cross, as an American, stands almost alone.

It is worth noting that the kind of engineering that Cross wanted to do was not at all common in the United States during the first quarter of the twentieth century. About 1910 the reknowned Göttingen mathematician Felix Klein visited the United States. On his return to Germany he gave a lecture in which he noted that the application of fundamental principles to engineering was not as important across the Atlantic as it was in Germany. A listener recalled Klein's remark: "The United States is such a rich country that it can afford to use twice as much material in a structure as is necessary. Why should a U.S. engineer be interested in a theory that tells him how to calculate exactly the amount of steel in a beam needed to carry a given load? If he's worried, all he needs to do is increase the amount of steel and make the beam stronger."[1]

At the same time that Klein was in the United States his equally distinguished colleague, Carl Runge, visited the country and made substantially the same observation. Theodore von Karman, the great aeronautical engineer, was at Göttingen at this time and wrote in his own autobiography that the United States was not well known as a nation able to produce learned men, nor was there much internal respect for them in the nation as a whole. By the time that von Karman himself first visited the United States, in 1926, matters had changed somewhat. Still, he was struck by the dominance of Europeans in departments of aeronautical engineering.[2] It is a measure of Cross's importance that he was widely recognized in Europe.

The genesis of this book occurred when I encountered a paper by Professor Holger Falter of the University of Stuttgart in *Nexus II: Architecture and Mathematics* (Fuccehio, Italy, 1998, ed. Kim Williams). The paper was entitled "The Influence of Mathematics on the Development of Structural Form." While I was in general agreement with Professor Falter's approach, I felt compelled to offer an emendation which has since been published in the *Nexus Network Journal,* Summer 2001.

Preparation of the emendation to Falter's article started me doing research ɔr this book. For one not skilled in the subtleties of structural engineering has been a daunting task. On the other hand, as an architectural historian

I have been dealing with buildings all my life and have been particularly attracted to the structural side of the discipline. In my last book, *Gateway Cities and Other Essays* (Ames, Iowa, 1989), I wrote about the advent of reinforced concrete in warehouses and industrial structures in the period 1900–1920. Subsequently I discussed Frank Lloyd Wright's use of the material in "Frank Lloyd Wright and the Concrete Slab and Column," *Journal of Architecture*, (1998): 315–45. Wright, I argued, used reinforced concrete in slab form in his early work. In the famous Johnson Wax Building of 1939 he conceived of his "lily pad" columns as three-hinged bents. Reinforced concrete when used as either a slab or a bent, however, displays a very different behavior from that exhibited when it is used on a continuous rigid frame, as in a skyscraper. With such a frame, the hinge problems cited by Professor Falter come into play.

All authorities agree that it was Hardy Cross who led the way in solving these problems. His initial breakthrough came in a famous paper of May 1930, published in the *Proceedings* of the American Society of Civil Engineers and reprinted in the *Transactions of the Society* in May 1932. As we shall see, its impact was tremendous. *Continuous Frames of Reinforced Concrete*, the textbook which Cross wrote with Newlon D. Morgan, was first published in 1932. By 1954 it had gone through thirteen printings without any revisions. It remained authoritative until the introduction of the digital computer.

Although this textbook was comprehensive for its time, the best way to understand the essential Cross is undoubtedly through the selection of papers made by Nathan M. Newmark, published by the University of Illinois Press in 1963. Newmark was Cross's best student and an eminent figure in his own right. Hence the second section of the present book is devoted almost entirely to his volume, which is entitled *Arches, Continuous Frames, Columns, and Conduits: Selected Papers of Hardy Cross*. For anyone who wants to make contact with the remarkable mind of Hardy Cross, Newmark's text is indispensable.

Despite the fact that Hardy Cross was an extremely articulate person, he did not leave a substantial collection of personal papers. During his lifetime he wrote thousands of letters, but only a few have survived. There is a small collection of documents at the University of Illinois. Through the courtesy of Professor Emeritus W. J. Hall I have been able to consult this resource.

Fortunately, we do have his academic records from Norfolk Academy, Hampden-Sydney, Massachusetts Institute of Technology (M.I.T.), and Harvard. We also have his teaching record at Brown University and valuable information about his colleagues at the University of Illinois. Finally we have the recollections of Robert C. Goodpasture and Thomas Golden and the invaluable notebooks of Thomas Kuesel. Mr. Goodpasture arrived in New

Haven in 1943 and took his B.S. in civil engineering in 1945. Because of commitments to the Navy, he remained in school until he had taken a master's degree in 1947. During this time Hardy Cross was chairman of the department and gave him much personal instruction. Goodpasture came to know Cross reasonably well, although he says Cross was "quite a private person." He was given access to Cross's personal papers, which were mostly manuscripts of speeches given before engineering societies. From these papers he put together the short book entitled *Engineers and Ivory Towers* (McGraw Hill, 1952). It was published a year after Cross's retirement from Yale and is an exceptionally valuable statement of Cross's views on engineering education, the role of engineering in society, and the problems of the engineering profession in the United States. On all of these topics Cross was an expert, and his remarks are well worth reading today.

I am deeply indebted to Mr. Goodpasture for his assistance in the preparation of this book. I am also indebted to the archivists of Norfolk Academy, Hampden-Sydney College, M.I.T., Harvard University, Brown University, the University of Illinois, and Yale University. My daughter-in-law, Ms. Brooksie Koopman, used her computer skills to point out to me that even today, more than two generations after its publication, Hardy Cross's method for calculating flow in networks is being used in colleges and universities across the country.

I want to pay tribute to my colleagues who gave me great assistance. Professor Emory Kemp, West Virginia University, not only encouraged me at every stage of this project but also wrote an appendix with worked examples of the slope deflection method, of column analogy, and of moment distribution. These will be of great interest to the technically minded reader. I have included it as Appendix C and cited it frequently in my text. It is only fair to note that Emory's definition of a statically indeterminate structure is slightly different from mine. I have used that of Cross, who stressed the idea of continuity. Emory's definition is more contemporary. His contribution has been immense and went far beyond conventional cooperation. It is a most unusual example of scholarly generosity.

The second colleague is Professor Emeritus Robert Darvas of the University of Michigan. Bob learned his structural engineering at Budapest in the years after the Second World War. When I mentioned Hardy Cross to him, his face lit up and he stated, "His method was absolutely worldwide." I have never forgotten his remark. Bob also read the manuscript and saved me from several egregious errors.

The impact of Hardy Cross was felt across Europe. Through the courtesy of Professor Jacques Heyman of Cambridge University I was able to study

European reactions to Cross's thought at the excellent library of the Institution of Civil Engineers of Great Britain on Great George Street, London. Mr. Mike Chrimes, the librarian there, was most helpful. Professor Heyman, an authority on the history of structural theory, gave me much encouragement with this project. I would like to acknowledge my indebtedness to the staff of the Public Library at Newport, Oregon. They inevitably handled my requests for scarce materials on inter-library loan with skill and cheerfulness. I am particularly grateful to Martha Hawkes.

Finally I am happy to acknowledge the assistance of the Graham Foundation in this project. Ms. Sandy Patton of the Taubman College of Architecture at the University of Michigan displayed unfailing good humor in reading my sometimes difficult handwriting and preparing the first version of this manuscript. Ms. Rose Reed of Newport, Oregon, helped in the later stages of manuscript preparation.

Any account of Hardy Cross's achievement must be based primarily on his own published work. Happily his personality was so strong that it comes through in almost everything he ever wrote. This quality is rare in engineering literature. It is a field in which writers strive for objectivity. Often it is difficult to ascertain anything about the character of a man or woman who has written an important work. With Cross, on the other hand, the intellectual integrity and enormous clarity come through in every line. Also evident is an exceedingly dry sense of humor. It is somehow not surprising that *Alice in Wonderland* was a favorite book of his. Cross knew so much about the theory of elasticity that he seems to have thought of himself as a character in Lewis Carroll's comic masterpiece. In the classroom there was nothing sweet about him. He was tough and demanding, direct and honest in his approach. In every respect he was a memorable personality.

In the United States structural engineers usually became famous in one of two ways. (1) They design and build some extremely important public structure such as the Brooklyn Bridge. Everyone knows the story of the Roeblings, father and son. The bridge has been celebrated in poetry and painting. It is one of the nation's greatest cultural artifacts. (2) They associate with prominent architects on some nationally significant building. Thus the name of Paul Mueller is forever tied to that of Frank Lloyd Wright for Mueller's contributions to Unity Temple and the Imperial Hotel in Tokyo. August Komendant could write a book entitled "Twenty Years with Architect Louis I. Kahn," detailing his contributions to the Salk Labs and the Kimbell Museum. Eero Saarinen relied on Fred Severud for help with the famous Ingalls Hockey Rink at Yale. Mies van Der Rohe had Frank Kornacker with him for the Seagram Building. The list of engineer/architect partnerships could be extended indefinitely.

Hardy Cross took neither of these routes. Instead, he was primarily a structural theorist, but his contribution to the building art was surely as great as that of any of those just mentioned. It was of critical importance to the most gifted architectural designers of his period.

In some ways the interpretation of Hardy Cross is made more difficult by the fact that he was essentially a nineteenth-century personality who lived on into the twentieth century but retained many nineteenth century attitudes. He grew up in a part of the world which was recovering from the effects of the Civil War. The virtues emphasized by his parents were hard work, economy, and careful husbandry. He had the most classical education imaginable, precisely the opposite of everything that John Dewey stood for. Cross had no desire to put his name on an important structure. He was skeptical of much university research, and in his own work he used very few footnotes: He did not need many because he was profoundly original. To the end of his days he had the manners and deportment of a Southern gentleman. It is not surprising that he did not get along with his university president whose emphasis was on new buildings and research grants.

The adage "no news is good news" epitomizes the additional difficulty that occurs in researching Hardy Cross. The topic is definitely success in engineering design, which doesn't tend to make headlines. In recent years a small but much needed literature on building failure has appeared. Mario Salvadori wrote an excellent book entitled *Why Buildings Fall Down*. Professor Henry Petroski of Duke University has contributed another good volume: *To Engineer is Human: The Role of Failure in Successful Design*. On The Learning Channel, television producers created an excellent series of programs on building failure. Those shows were the basis for *Collapse!*, by Philip Wearne, another good book. These volumes set forth many reasons for building failure. Some collapses can be traced to shoddy workmanship. Others are the fault of misplaced faith in a structural member, such as the eyebar in the suspension bridge over the Ohio at Point Pleasant (1971). For some, like the infamous walkways of the Hyatt Regency Hotel in Kansas City in 1981, the structural engineers were clearly responsible. Broadly speaking, earthquakes have been probably the greatest single cause of collapse. But none of these collapses were attributable to a failure in the calculation of a building frame, and this is the study to which Hardy Cross addressed himself.

The most disastrous collapse in American history was that of the New Orleans levees at the time of Hurricane Katrina in September 2005. While investigations are ongoing, it is already clear that a major difficulty was poor construction of these levees. The reader will find an account of a remarkably similar problem on pages 58–59, which deal with the building settlement at the Charity Hospital. Those difficulties of 1939 foreshadowed the more recent collapse.

In writing this book I have been conscious that I was dealing with a technical subject and that some potential readers might be put off by the terminology which I have inevitably used. I have therefore included a glossary to explain the technical terms. My model has been the late Mario Salvadori, who wrote an excellent book, *Structure in Architecture*, without a single equation. I have not been able to emulate Salvadori completely, but I think I have come close. (Professor Kemp has supplied the equations in Appendix C.) In this connection I should note that there are various ways of looking at an important building and that the way of the structural engineer is an important, if often overlooked, method.

The ordinary tourist who admires a capital on a Greek column will be conscious of its beauty. If the eye is trained, the admirer may well be aware of the complex mathematics embodied in the capitals. But it is fair to say that only a rare visitor will think of the colonnade as a statically determinate system. Similarly the tourist will be overwhelmed by the soaring spaces of a Gothic cathedral, may respond to its assemblage of sculpture and stained glass, and may well understand the function of the flying buttresses. But unless this tourist is trained as a civil engineer, the response to them will not be as " . . . quite definitely columns under flexure." To describe the achievement of Hardy Cross it is necessary to use words like "moment" and "shear." I hope that readers will not be put off by these words and will refer frequently to the glossary.

One story about success in structural engineering is worth telling for the light that it throws upon Hardy Cross. In 1950 his prize pupil, Nathan Newmark, was invited to be a consultant on the Latino Americana Tower in Mexico City. The project would involve reinforced concrete, foundations, steel frame connections, and factors related to earthquake resistance. Newmark had an opportunity to use all the accumulated knowledge of his teachers at Illinois.

The problem was difficult. The tower was to be located in the heart of an area plagued by earthquakes. This area had soil unimaginably poor for the erection of a tall building. Two of Newmark's former students, Leonardo Zeevaert and Emilio Rosenblueth, worked with him on the job. When Newmark arrived to discuss it with the two young Mexican engineers, the steel for a twenty-story building had already been purchased and was on the building site. The three concluded that the proposed structure would be too weak for the earth motions it might have to withstand.

Newmark believed that with a better design the building could go as high as forty-three stories. So designing began. The first problem was the ground itself; watery clay extended down to a depth of a hundred feet or more. Below this there were layers of fine sand, gravel, and more clay. An historian of the building relates:

> Zeevaert and Newmark designed the Tower to stand on 361 one-hundred-foot concrete pilings that were driven down to the level of the sand, 117 feet below street level. The foundation was planned to behave as a floating box, sitting on these roots in such a way that the load of the building was carried by the piles to the thin, hard stratum of sand. The building was to be as light and flexible as possible. The tops and bottoms of all the windows were anchored with single bolts, and small spaces were left between the tops of partitions and the ceilings. The building was designed to withstand earthquake tremors three times stronger than any previously recorded in Mexico City.
>
> As completed, the Tower stood 43 stories high, with a 138–foot television antenna towering more than 600 feet above the city. Instruments were mounted at various elevations to record motions of the floor due to winds or earthquakes. Newmark's part of the work was concluded in 1954, and the structure became another part of his past.
>
> After the earthquake of 1957, engineers went into the Tower and examined the recording instruments. The top of the building, they learned, had whipped back and forth approximately a foot. Cushioned by its special foundation, the gigantic structure had undergone almost exactly the forces it had been designed to withstand. Many sightseers, walking or driving through Mexico City after the earthquake, were amazed to see the city's tallest building standing undamaged.[3]

Hardy Cross would not have been surprised. He might well have remarked that Newmark was a man to be trusted. The story, of course, is dramatic. Less dramatic but more important were the thousands of buildings whose frames were successfully computed by the "moment distribution method." They are the real monuments to Hardy Cross.

1. The Preparatory Years, 1885–1921

> I have fantasized that were I ever to be in charge of making arrangements for an imaginary cocktail party for the intellectual stars of history, among my invitees would surely have been Hardy Cross. Why? Perhaps because his iterative approach, rooted in his philosophical thinking, found a solution to what had been the prior mystery of indeterminate structures, and it seemed to me to match the intellectual musings of Galileo, Newton, Einstein, etc. How do things work? What is important? Can the solution be described in mathematics within a range to satisfy observational success? I suspect reinforced concrete became his baby because of the combined inexactitude of the working together of hardened cement and reinforcing steel when coupled with the minuscule variation from an absolutely precise computation of stresses by his method over prior methods.
> —Tom Golden, a student of Hardy Cross, in a letter to the author, 10 March 02

At first glance Tidewater, Virginia is an unlikely milieu for the development of eminent academics and world-famous engineers, but it is the setting for the first part of this book. No part of the area is more southern than Nansemond County, which was merged with Suffolk County in 1974. It attracted European settlers as early as 1635. Fortuitously, one might say, the soil was not much good for the growing of tobacco, which was the curse of Virginia agriculture for so many years. On the other hand, it proved ideal for the cultivation of peanuts, introduced by the slave traders. It also proved an excellent place to raise the famous Virginia hogs. Hams from nearby Smithfield are famous to this day. In sum, it was a rich agricultural area, and as might have been expected, it was extremely loyal to the Confederacy. The county raised nine companies of infantry for the Army of Northern Virginia. One of the riflemen was Thomas Hardy Cross, heir to a plantation near the Great Dismal Swamp. The place was called "Farmer's Delight." An ancestor of Thomas Hardy Cross had purchased it from Thomas Swepson in the late eighteenth century.[1]

Thomas Hardy Cross was born October 12, 1841, and grew up at "Farmer's Delight." He attended the University of Virginia in the 1858–59 academic year,

Thomas Hardy Cross in Confederate Uniform. Cross Archives, University of Illinois.

and two years later went to war. Though the only battle at which he is known to have been present was at Cumberland Church near Farmville, he must have had a hard war. Family lore has it that Cross was with Lee at Appomattox. One source notes that he went to South America after the war. Perhaps he was one of those Confederate veterans who thought of trying for a fresh start in life in that region. In any event, he came back to "Farmer's Delight" and in 1870 sold off part of the land, an action which suggests that, like so many southerners in those years, he was land poor. In the early 1870s his existence must have resembled that portrayed in Ellen Glasgow's *Barren Ground*. It would have been one of unremitting toil. Happily his exertions seem to have been rewarded. The Crosses were successful farmers and solid members of the middle class. Late in life Hardy Cross remarked that his father was "active in politics."

Virginia in the years from 1865 to 1900 was in most regards an extremely conservative place. After the upheaval of Reconstruction, the pre–Civil War elites successfully reasserted their control over the affairs of the state. It was the view of these elites, headed by W. T. Mahone, one of the former com-

Thomas Hardy Cross at the Battlefield of Cumberland Church in 1902. Cross Archives, University of Illinois.

manders of Thomas Hardy Cross in the Army of Northern Virginia, that the way out of the state's problems was to be found through tight fiscal management by the state government and railroad development by private interests. Mahone's program was to "readjust" the state debt, which had grown prodigiously during Reconstruction. He repealed the poll tax and passed legislation to improve the schools, but his overall approach was conservative. This approach was taken over first by Carter Glass and more recently by Harry Byrd. Mahone, Glass, and Byrd were good examples of the conservative Southern Democrat who believed that the way ahead meant adherence to time-honored principles.

In 1907, a chronicler of the recent past of Nansemond County wrote, "the county was in the possession of the Federals for nearly three years and her resources were exhausted by the support of an immense army quartered in her midst."[2] Recovery from this state of affairs was surprisingly rapid. It was based on the excellent productivity of the soil, on which corn, cotton, and most importantly, peanuts were visibly and successfully cultivated.

There were in the county already seven large factories for the cleaning and shelling of peanuts. Their output had an annual value of over $3 million. These factories shipped their product to market over a network of railways which developed during these years of progress. No less than six railroads passed through Suffolk, the principal town of the county. They included the Norfolk and Western, the Atlantic Coast Line, and the Southern. All these enterprises needed bridges, and these structures may well have been noticed by the younger of the two boys growing up on the Cross plantation.

We do not know exactly when Thomas Hardy Cross met Eleanor Elizabeth Wright of Wright's Point Farm near Smithfield, nor do we know much about the lady herself. Circumstances suggest that she was one of those capable women who took on great responsibilities during the Civil War. She married Thomas Hardy Cross on January 13, 1879. Her age at marriage was given as thirty-seven and his as thirty-eight, so they would both have experienced the Civil War and Reconstruction. Their first son, Thomas Peete Cross, was born on December 8, 1879. Their second child, Hardy Cross, arrived on February 10, 1885. Perhaps the most important problem this couple confronted was the lack of proper schooling for two extremely bright boys. This part of Virginia was and is extremely rural. Opportunities for a systematic education were non-existent. Tom Peete Cross's application for Harvard University indicates some home tutoring. Mr. and Mrs. Cross must have recognized that they had two exceptional youngsters on their hands. Their solution was to send the boys away to Norfolk Academy, the nearest independent preparatory school. Tom Peete went in 1891. He did exceedingly well, finishing at the top of his class.

Norfolk Academy had a major effect on Tom Peete Cross and his brother Hardy, who enrolled in 1893. The school, still thriving today, has a history going back to 1786, when, though the town was still in ashes, the inhabitants impoverished, and taxes too high, the council appointed a committee to contract for building a free school. The building in which young Hardy Cross went to school was a handsome Greek Revival edifice erected in 1840 to a design by Thomas U. Walter. It was modeled on the Athenian Theseum and may well have been Hardy Cross's first contact with sophisticated architecture. During the Civil War, the school was taken over by the Union Army and used as a hospital, but in the autumn of 1865, the building was renovated, new equipment was secured, and new teachers were engaged. The first post-war students were accepted in October. In 1877 the City Council attempted to take it over. This attempt was rebuffed, and it entered a new period of usefulness as a private school. In 1882 Robert W. Tunstall, a University of Virginia graduate, was elected principal. He brought the curriculum up to date, engaged well-trained teachers, and secured adequate equipment. Tunstall, who was evidently a man of great force of character, was headmaster of the school when Hardy Cross arrived.[3]

What sort of place was this school where Hardy Cross was to spend the next six years of his life? In the 1897–98 school year, when he was in the tenth grade, it had a total enrollment of 117. The youngest boy in the school was ten, the oldest twenty-one. Including Principal Tunstall, who taught Latin, Greek, and mathematics, it had a faculty of five men. These individuals had degrees from the University of Virginia, Johns Hopkins University, and the University of North Carolina. A description of the school declares:

> The Academy is designed to give to boys and young men a sound training in the subjects of secondary instruction. There are six Forms, of which the four highest constitute the Upper School. The First Form (the lowest) is divided into two sections, and occupies two years.
>
> The work of the First and Second Forms is directed mainly toward a careful training in the elementary English branches. Elementary Latin, however, is begun in the lowest Form, and those who intend ever to study any foreign language are urged to begin Latin as early as possible.
>
> In the Third Form the student enters upon secondary work in preparation for college or the practical duties of life.
>
> Constant review work characterizes the entire course, so that any failure that may occur in the effort to secure thoroughness of results cannot justly be charged against the school.

Courses of Study

> The only optional study in the Lower School is Latin, but in every instance the pupil, in these elementary grades, is advised to study Latin.
>
> In the Third and higher Forms, the student should, under the guidance of his parents and the Principal, make such a selection of studies as shall best conform to his future needs. The mistake most commonly made in this respect is made by fathers and mothers who permit their sons to undertake too little. "He who does not know foreign languages, knows nothing of his own." Every boy in the school should take at least one foreign language throughout his course. Most imperative of all is that he take the full course in English, including reading and spelling.[4]

The insistence on Latin is at first glance reminiscent of the classical traditions characteristic of the Old South, but here it seems to have been seen more as an aid to understanding the structure of French and German. The Latin authors studied at the Academy occur in the same sequence in which this writer encountered them in late 1930s: Caesar in the tenth grade, Cicero in the eleventh, and Virgil in the senior year. At every stage these authors were read in conjunction with textbooks on grammar and composition. The literature on the Academy says nothing about discipline, but one certainly has the impression that it was strict. The grading system was numerical. A passing mark was 70; a distinction mark was 80; a prize mark was 92. The school had ten scholarships awarded annually to " ... such deserving boys of the city as are unable to pay the tuition fees." Tuition and board, including fuel, light, and laundry per session, was $275, payable one half at the time of entrance and the remainder February 1st. There were discounts to parents sending more than one son, and also to clergymen and teachers. The only extra-curricular activity was a monthly magazine published by the students. It was entitled "The Orange and White" and was thought of as a useful adjunct to the department of composition and rhetoric. Headmaster Tunstall took a limited number of boarders in his home. These boys had to provide in advance for books and stationery and furnish their own blankets and towels with each article properly marked. It is probable that the Cross brothers boarded with Headmaster Tunstall unless there were relatives in Norfolk. So far we have been unable to discover any family in the city.

Tom Peete and Hardy Cross were brilliant students. Tom was a prizeman with an average of 92 in the ninth grade, 93 in the tenth, and 94 in the eleventh grade. Hardy did even better. For the 1898–99 school year he had 100 in Plane and Solid Geometry. One wonders if he missed any problems during the year. In later life he was to argue that geometry was an essential discipline for the structural engineer. In Latin he scored 93, in French 94, in German 96, in History 97, and in English 96. His grades in reading and spelling were 97 and 96 respectively. His overall standing was 97, which placed him second by three-tenths of a point in a class of fifteen.[5] His textbooks were not snaps. They included Phillips and Fisher's *Plane and Solid Geometry* and Gildersleave's *Latin Grammar*. The English reading list for that year included: *Last of the Mohicans, The Sir Roger de Coverly Papers,* and *Silas Marner*. In French, young Cross used Whitney's *Grammar* and read Dumas' *L'Evasion du Duc de Beaufort,* Racine's *Athalie,* and Dumas *Les Trois Mousquetaires*. In German, he used Whitney's *Brief Grammar* and read Schiller's *Der Neffe als Onkel* and Hauff's *Das Bild des Kaiser*. In the eighth grade Cross most likely had a brief encounter with physics, but he had no real acquaintance with that subject until he reached college. Hardy Cross never graduated from Norfolk Academy in the sense that he did not complete the sixth form (twelfth grade). His family and his headmaster simply thought that he had advanced far enough in his preparatory studies so that he was ready for college.

In his old age, Cross told an interviewer, "The best education I ever had was at Norfolk Academy under Bob Tunstall. The curriculum was simple and there wasn't much science, but we learned there what was more important than the subjects themselves: how to think and study."

In the same interview Hardy Cross commented on the excellence of the mathematics curriculum. It included conic sections, which would today be termed analytic geometry. The school had, he said, higher academic standards than any other in the area, and he added:

> In 40 or 50 years the facts about a subject are likely to either be discarded or else thrown into a new light. In education today much time is spent on formal courses which do the student almost no good when he completes them. Today there is far too much emphasis on formal education, on completing certain courses, attaining certain degrees.
>
> Before the Civil War there was little formal education, but there were many brilliant, well read, learned people. Today much is taught that is not essential. Many teachers, especially in college, lack vision, enthusiasm and just go through the formalities of teaching. Thus a person can have much formal education and still be poorly educated.
>
> The important thing is not to assimilate many facts, but to learn to collect and evaluate facts; not to know what is thought about a subject but to be able to think and analyze. The purpose of an education is not to state what is so, but to show how best to determine what is so. Because it taught me these things, my education at Norfolk Academy has been more valuable to me than any other.[6]

We believe that in certain respects the milieu of the post–Civil War South had an effect on Hardy Cross's career and on his approach to the problems of structural engineering.

If one wanted to briefly characterize Cross's professional objectives, one would have to say that the man sought to revolutionize the discipline of structural engineering. He achieved his end, but his first step was a restatement of physical principles first enunciated by James Clerk Maxwell in 1864. We shall take up his paper on the subject presently. But it is notable that Cross firmly believed that such a restatement could often be as important as a new discovery. All his life he was on the lookout for specious novelty, not only in civil engineering but in all the arts and sciences. Scholarship, research, productive investigation, he wrote, are often the last refuge of academic charlatans. This attitude, combined with his own tremendous learning and enormous honesty, undoubtedly made him a trial to many administrators.

In the fall of 1899 Hardy Cross enrolled at Hampden-Sydney College in central Virginia. This institution, about one hundred miles south of Charlottesville, dated back to 1776 and was originally a Presbyterian foundation. It is said that the original curriculum reflected the ideas of John Knox! The

Hardy Cross as a student at Hampden-Sydney.
Cross Archives, University of Illinois.

place was one of several southern offshoots of Princeton, and it had a strong appeal to the sons of that Scotch-Irish stock which was so important in peopling the great Valley of Virginia. Hampden-Sydney, named after two important seventeenth-century English Puritans, had its ups and downs during the nineteenth century, but by the time Hardy Cross arrived, it was a thriving place. The curriculum stressed the traditional liberal arts disciplines: the Classics, English literature, mathematics, the physical sciences, and religion and theology. The college emphasized teaching and public service. In 1845 Henry Howe reported that it had sent more teachers into the southern part of the United States than any other college. This emphasis must have been congenial for Hardy Cross, who was nothing if not a born teacher.

Hampden-Sydney at the beginning of the twentieth century required a formidable series of courses for the B.A. and B.S. The B.A. degree was conferred on those who completed *all* the studies of the freshman and sophomore classes and *all* of the compulsory studies of the junior and senior classes: Moral Philosophy (including Bible Studies), Physics, Chemistry, Latin or Greek,

English, and Political Science. There were three elective hours in the junior year and eight elective hours in the senior year, and there was some possibility of substituting French and German for freshman and sophomore Latin and Greek. Mathematics was particularly strong, with offerings up through differential and integral calculus. A curious feature of the latter program was the continuing use of C. I. Venable's *Notes on Solid Geometry*, a textbook first published in 1845. There was no suggestion anywhere that the program was a preparation for the world of business or for making money. These were not its objectives. It was a liberal arts institution with an emphasis on pre-professional education, especially for the theological seminary. And for Hardy Cross, the courses in mathematics, and in English, were a superb preparation for an engineering education. Mathematics was a tool which he would use every day of his professional career, and a lively and provocative English style became one of the hallmarks of his numerous papers.

One particular custom at Hampden-Sydney may have been important for Cross. Among the great wonders of the age was the High Bridge at Farmville, which carries the main line of the Norfolk and Western Railroad (now Chesapeake Seaboard Express) from the coal fields of West Virginia to Norfolk. This bridge figured crucially in the last days of the Civil War and was the reason that the Reconstruction government moved the county seat from Worsham (two miles east from Hampden-Sydney) to Farmville in the early 1870s. It was garrisoned in World War I and World War II, and it still rates as a remarkably long non-suspension bridge. In Cross's day it was common for fraternities (the Cross brothers belonged to Kappa Alpha) to have picnics on a bluff from which the bridge—and, of course, the long trains crossing it—could be viewed. And at the turn of the century there was much ambitious talk of building great bridges to link Portsmouth, Newport News, and Norfolk. Hardy Cross may have looked at this remarkable bridge and pondered his future career.

Cross took his B.A. in 1902, and was valedictorian of his class. As was the custom for valedictorians in those days, he stayed on for a year as College Fellow, helping to instruct the sub-freshman class—those boys needing extra preparation for the entrance exams. Concerning this year, he said, "I taught a little of everything including English, mathematics, and science." In 1903 the College granted him a B.S. degree. Aside from his academic work, he was editor-in-chief of *Magazine*, on the staff of the *Kaleidoscope* (the yearbook), treasurer of his class in the second term of the 1901–2 year, and Final Seminar Orator. His physics professor, Dr. J. H. C. Bagby, said of him, "In a long experience with college men, I have never met Hardy Cross's superior in ability to think straight, in power of keen analysis, and in bril-

Bridge at Farmville. From Edward Berger, *Album of Virginia*. Courtesy of the Library of Virginia.

liant deduction." Everything suggests that Cross made a strong impact on Hampden-Sydney during his college days.[7]

In the 1890s, Hardy Cross's parents took an extraordinary step. They sold the family plantation and moved to Hampden-Sydney in January, 1892, to open a small store and a boarding house for students. Thomas Hardy Cross became postmaster. Obviously the Cross parents wanted to give their extremely capable sons the best possible start in life. The move indicates a strong family unit and also suggests that Eleanor Cross had substantial managerial abilities. She had to run the boarding house and keep its accounts while her husband ran the store and presided at the post office. This managerial ability seems to have been common among southern women, but was often well camouflaged. Happily, Mrs. Cross lived until 1926, by which date her two sons were well established in their careers. Perhaps it is significant that Hardy Cross and his wife are both buried in the Wright family plot at Smithfield. Both boys were prodigies and more than justified their parents' confidence.

While this work is about Hardy Cross, it is relevant to point out that his older brother also had a successful academic career. After taking a B.A. in 1899 and a B.S. in 1900 at Hampden-Sydney, Tom Peete Cross returned to Norfolk Academy to teach modern languages for five years, from 1900 to 1905. He took a decisive step toward moving away from the South when he took an M.A. from Harvard in English in 1906 and a Ph.D. from the same institution in 1909. Subsequently he taught at Sweetbriar College from 1911 to 1912, the University of North Carolina from 1912 to 1913, and then at the University of Chicago, where he remained for the rest of his academic career. His specialty was Celtic languages. He was chairman of the Department of Comparative Literature until 1930, a member of the Council of Irish Texts Society, a fellow of the American Irish Historical Society, and at Chicago, President of the Quadrangle Club in the 1944–45 school year. Tom Peete Cross was the author of numerous scholarly works on English and Irish literature and always extremely loyal to Hampden-Sydney. To that college he bequeathed the (very scant) royalties of his monumental tome on Celtic mythology.

In these days when the dreaded sibling rivalry is so much discussed, it is worthwhile to note a remarkable parallelism in the careers of the two brothers. Both young men made outstanding records at Norfolk Academy. Both went on to do equally well at Hampden-Sydney. Both returned to Norfolk Academy to teach. Both decided to go to New England for graduate training. It is reasonable to suspect that Tom Peete Cross may well have played a role in showing his younger brother the ropes in the unfamiliar milieu of Cambridge and Boston. After their graduate years their careers parted, but during the long period when Hardy was at the University of Illinois in Urbana-Champaign, Tom Peete Cross was a major figure at the University of Chicago. They may have seen a good deal of each other. It is almost as if the older brother served as a model for the younger.

After Hardy Cross completed his years at Hampden-Sydney, he returned to Norfolk Academy to join the faculty. He spent three years at his old school teaching mathematics and English. Cross was glad to return to Norfolk, but after three years he was restless. In 1937 Cross declared that the reason for his decision to enter the engineering profession had been extremely simple. None of his family had ever been engineers, but he could see little future in teaching liberal arts subjects in preparatory schools. Once in the engineering profession, he found that it intrigued him, and he made a career of it.

In 1906 he left Virginia for the first time and entered the Massachusetts Institute of Technology. M.I.T. recognized his degrees from Hampden-Sydney and awarded him credit for some of his work taken there. He took two years to complete his work for the bachelor's degree. His performance was strong but not outstanding. In integral calculus he took a "C," and in physics he was noted "satisfactory." His best grades were in applied mechanics and hydraulic measurements. He also did well in bridge design. In his second year he was awarded a scholarship of $50 from the Vose Fund. His senior thesis was on "Experimental Studies of the Lateral Thrust and Angle of Internal Friction of Soils." He took his degree in 1908, but did not have the record of a man who was destined to make a major contribution to structural engineering.[8] His only extracurricular activity was membership in the Southern Club.

In later years Cross would be primarily concerned with reinforced concrete. Since Boston was the first urban environment which he experienced, it is worth pausing to examine what he might have seen there. It is commonly known that the first reinforced concrete structure in the United States was the William Ward house, built in Port Chester, New York, in 1870. In 1890, Ernest L. Ransome built the two-story Leland Stanford Museum in Palo Alto, California, and in 1897–98 the factory for the Pacific Borax Co. The turn of the century was a period of intense interest in reinforced concrete. Thomas Alva Edison thought that it had a great future and built a few houses of the material. By the time Cross started teaching at Brown in 1911 there was substantial literature on the material, but there were also innumerable unanswered questions.

In Boston itself there were few, if any, concrete buildings, but in the suburb of Beverly, Ransome had erected a stunning series of factories for the United Shoe Machinery Company in 1903–6. These bold structures of reinforced concrete were built with a system closely resembling that of François Hennebique. Hennebique devised a method of construction used widely in reinforced concrete. Beginning with floor slabs in 1879, he progressed to a complete building system which he patented in 1892. It featured structural beams of concrete reinforced with stirrups and designed to resist the tensile forces against which ordinary concrete was weak. For the most part, American builders and architects at this time were interested in flat-slab construction, but Cross left that aspect of the concrete problem to his gifted colleague, Harald Westergaard. His own interest was in continuous frames. The most important high-rise example of reinforced concrete construction in the United Sates during this period was the Ingalls building in Cincinnati (1903), a city which Cross is not known to have visited. Ransome's structures for the shoe machinery company might, on the other hand, have been provocative.

In the fall of 1908, Cross, M.I.T. diploma in hand, took a job in the bridge department of the Missouri Pacific Railroad in St. Louis. It was his first experience of the West, and he must have found it exciting. Although by 1908, St. Louis had lost out to Chicago as the premier midwestern railroad entrepot, the city had immense commercial and cultural vitality. For a young engineer there was the famous Captain James B. Eads bridge across the Mississippi, (1872–73). There were two fine steel-framed skyscrapers by Adler and Sullivan: the Wainwright Building (1891) and the Union Trust (1892). In an older tradition there were numerous good cast-iron buildings on the riverfront.

Once again, however, there was nothing much in reinforced concrete. The great warehouses of Deere and Co. (Oscar Eckerman, 1904) were framed in timber. As Reyner Banham pointed out, it is an oddity that the most important American developments in reinforced concrete took place in provincial centers like Minneapolis and Buffalo.[9]

In 1909, after a year with the Missouri Pacific Bridge Department, Cross returned to teach again at Norfolk Academy. His departure from his post in 1910 marked a firm commitment to the profession of civil engineering. On October 4, 1909, he wrote to the Dean of the Graduate School of Applied Science at Harvard University that he was thinking of taking advanced work in Civil Engineering and Applied Mathematics during the academic year 1910–11 asking for a catalogue "or curricular information bearing on the subject." On October 8, 1909, the dean replied to his letter and enclosed a pamphlet. The following autumn Cross enrolled at Harvard and in June, 1911, took the M.C.E. degree from that institution. His grades were better than those at M.I.T. They were in fact, good enough for him to be offered a position as Assistant Professor of Civil Engineering at Brown University to commence that fall.[10]

So Hardy Cross moved to Providence, Rhode Island, as the fourth man in a small department of Civil Engineering. As might be expected, the entire School of Engineering was small. It featured curricula in civil, mechanical, and electrical engineering. Of these, the program in civil engineering was probably the strongest. A glance at Cross's teaching assignments during his first year, 1911–12, is instructive:

4. Surveying, Advanced Field Work

   Use of stadia, plane table, and barometer; precise base-line measurement with steel tape; triangulation; topography. Methods as in 1, 2.

   Professors Hill and Cross, Messrs. Hutchins and Long.

   One hour. Spring recess. Elective for students who have credit for 1 and are taking 2.

15. Framed Structures

    The standard forms of simple roof trusses; calculation by analytical methods of stresses due to dead, snow, and wind loads; calculation by analytical methods of stresses in Pratt, Howe, Warren, and parabolic trusses due to dead, wind, uniform live, excess panel live, road roller, and locomotive wheel loads; miscellaneous trusses, skew bridges, cranes, trussed bents, and towers; general discussion of continuous, cantilever suspension bridges and arched ribs, and continuous and non-continuous center-bearing swing bridges; deflection of framed structures; theorem of least work.

Lectures on American and English shop practice; the aesthetic design of bridges; design of standpipes and elevated tanks; the American methods of erection of bridges and structures. Lectures, recitations, and seminary work. Professor Cross.

Three hours. First semester. Elective for students who have credit for Mechanics 2. Mon., Tu., Th., at 8.

17. Elements of Structural Design

Center of gravity and moment of inertia of combinations of structural shapes; riveted joints, tension members, compression members; combined direct and bending stresses, secondary stresses, columns under direct and eccentric loading; wooden structures; plate girder, pin-connected, and riveted Pratt truss bridges; construction of viaducts and elevated railroads; steel mill-building and high-building construction; comparison of standard structural specifications. Lectures, recitations, and computations. Professor Cross, Messrs. Hutchins and Long.

Two hours. First semester. Elective for students who have credit for Mechanics 2. Tu., 9.20 to 11.20.

19, 20. Bridge and Roof Design and Graphic Statics

Graphic determination of center of gravity of combinations of structural shapes, of moments in beams and stresses in simple roof trusses; determination of shears, moments, and stresses in plate girders and bridge trusses by the use of influence lines; complete design, with general drawings, of a roof truss; revision and discussion of checked drawings; design, with general drawings, of a plate girder highway bridge, and of a Pratt truss railroad bridge. Lectures, drawing, computations, inspection trips, and seminary work. Professor Cross, Messrs. Hutchins and Long.

Three hours. One hour of recitation and four hours of drafting, or six hours of drafting through the year. Elective for students who have credit for Mechanical Drawing 4, and have credit for or are talking 15 and 17. First semester. Tu., 11.20 to 1.20; 2.20 to 6.20. Second semester, Tu., 11.20 to 1.20; 2.20 to 6.20.

22. Reinforced Concrete and Masonry Structures

Theory and design of reinforced concrete beams, columns, retaining walls, dams, foundations, bridges, and culverts; systems of reinforcements; methods of construction. Stone masonry; static and elastic theory of the masonry arch; theory and design of masonry piers, retaining walls, bridge abutments, and high masonry dams; complete design and drawings, of a reinforced concrete highway bridge; a reinforced concrete retaining wall; a voussoir stone arch, abutments, and wing walls; and a high masonry dam. Lectures, recitations, drafting, and computations. Professor Cross, Messrs. Hutchins and Long. Five hours. Two hours of recitation and six hours of drafting. Second semester. Elective for students who have credit for Mechanics 2 and Mechanical Drawing 4. Mon. at 10.20; Th., 10.20 to 1.20; 2.20 to 6.20.

24. Foundations

Soundings and borings, foundation-beds, deposition of concrete under water, timber foundations, coffer-dams, open caissons, Cushing cylinder piers, pile formulae, driving and foundations, pneumatic caissons of wood and steel, open dredging, the Poetsch freezing process, foundations in quicksand, foundations for high buildings.

Lectures, recitations, and seminary work. Professor Cross. One hour. Second semester. Elective for students who have credit for Mechanics 2. Mon. at 9.20.

26. Law of Contracts

Essential elements of contract; the parties to a contract, considerations, bids and bidders; work for private parties and public work. Employment of engineers, liability of the engineer as a professional man; when his functions are judicial; when he is a public officer. Engineering specifications and accompanying documents. Lectures and discussion of written decisions on typical cases assigned each week.

Professor Cross. One hour. Second semester. Required for all candidates for the degree of Bachelor of Science. Th. At 9.20.

28. Engineering Jurisprudence

Patents and patent laws including inventions, novelty, utility, abandonment, applications, letters patent, reissues, infringements, damages, injunctions, and profits; real property, water rights, ownership, rights of way, boundaries, incorporeal rights, and franchises. Lectures, recitations, and seminary work. Professor Cross. One hour. Second semester. Elective for Juniors and Seniors. Tu. At 9.20.

* Courses 19, 20 and 22 may not be elected separately.[11]

Obviously Hardy Cross was a hardworking young academic. Some of his courses were taught with colleagues, but for most, he alone was responsible. Despite his experience with the Missouri Pacific Railroad, he gave no courses in railroad engineering, an important subject at the time, nor did he give any courses in hydraulic engineering. There is a certain irony in this. During his lifetime, Cross achieved great fame for his contributions to the theory of elasticity, a subject which he introduced into the curriculum at Brown in 1913. Forty years after his death in 1959, his work on flow in conduits or conductors in a paper of 1936 is a fundamental technique for hydraulic (and electrical) engineers. Professor Hill, Cross's department chair, was wise in giving him an opportunity to develop his interest in structural theory and in steel and concrete, which Cross taught in the 1914–15 year and regularly thereafter. And it is impressive that in 1917 he gave a course on Construction in Steel and Concrete, "primarily for those who will become purchasers rather than designers of engineering structures. Principles and methods of design are explained and the process of construction outlined from foundation to complete structure, the functions of each worker from laborer to purchaser being indicated. Practice in reading and preparing structural drawings." This description reminds one of the "Architecture for Non-Architects" courses which have proliferated in architectural schools since 1950. It is, after all, tremendously important to educate those who will be the patrons of architecture. Here again Cross was ahead of his time.

The years at Brown were a rewarding period for Hardy Cross. He obtained a broad view of the major problems in civil engineering education, and was able to identify those on which he wanted to work. As early as 1914 he was teaching a course on statically determinate structures and paying attention to some of the more common cases of statical indeterminacy. He was able to experiment with both graphic and mathematical techniques of analysis, and he came to see the value of both—though he had a healthy skepticism of elaborate mathematical solutions. For Cross, the most appropriate method was often that which was simple and direct.

From all accounts Hardy Cross had a pleasant personality and was a great gentleman. Nonetheless in 1918 he resigned from Brown to go into private practice. There were probably several reasons for this decision. In those days academic promotion was slow, especially at Ivy League schools. Seven years had passed since his appointment, and he was still an Assistant Professor. Secondly, he had decided to marry, and he needed more money.[12] Finally, he may have been a bit uneasy over the position of civil engineering at what was still basically a liberal arts university. So he bade farewell to Brown and joined the office of Charles T. Main, a prominent civil engineer in Boston.

Charles T. Main (1856–1943) was an expert on power problems, especially those of the New England textile industry. He knew both steam and electrical power well and was skilled in mill construction. Trained as a mechanical engineer, Main was elected a Fellow of the American Society of Mechanical Engineers in 1936 and also served a term as President of the Boston Society of Civil Engineers. While holding this position he drafted the first code of ethics adapted by an engineering society. His interests were broad, and he had high professional standards. The author of a brief biography remarks: "Engineering was his natural bent. He followed the principles of his forebears in working hard and long. One could not escape the impression that he was frictionless. Perhaps that accounts for his long and useful life. His mind worked as a philosopher's on technical subjects, resulting in much original thinking and pioneering in fields then little explored. The study of industrial values, and in finding reasonable means of measuring them so the businessmen could use them especially interested him."[13]

Some of this language could easily have applied to Hardy Cross as well. He, too, had a "philosophical mind" and did original thinking in unexplored fields. On the other hand, he seems never to have been much interested in the business side of engineering—although he very much enjoyed the consultancies which brought him into contact with the business world. Since the records of the firm of Charles T. Main have been lost, we do not know the projects on which he worked, but it is fair to assume that he learned a great deal. By the time he went to the University of Illinois in 1921, he had evidently selected the problems on which he wanted to spend his career.

2. The Creative Years at Illinois, 1921–37

> All great discoveries carry with them the indelible mark of poetic thought. It is necessary to be a poet to create.
> —E. M. Bataille

What is a statically indeterminate structure? To understand Hardy Cross's achievement, we must be clear about this question. When a beam rests on two supports it is subject to bending and is said to be *determinate*. The span can be considered in isolation. The exterior colonnade of a Greek temple is a wonderful example of a determinate system in stone. The action of external forces is entirely countered by the forces between every two columns of the series. But if the stone block were to extend over three or more supporting columns, the system would be classified as *statically indeterminate*. No span could be isolated because each would be influenced by the action of external and internal forces of the adjacent spans. This case is much more complex theoretically and mathematically, and it is much more common, especially with an industrial civilization which desires to construct metal frameworks for railway bridges across several supports and metal frames for tall buildings. In both building types the rigid connections between beams and columns create complex interactions. Hardy Cross wrote that the Baths of Caracalla were continuous (they were done in Roman concrete). Most medieval structures in stone were also continuous. A gothic cathedral, then, had structural resemblances to the frame of a modern skyscraper. In both building types the feature leading to indeterminacy was *continuity*.

About 1850 a French engineer, B. P. E. Clapeyron, who worked mostly for railway companies, evolved a new method to solve the problem of the statically indeterminate structure. He called it the Theorem of the Three Moments, and it was useful in the design of bridges. For the next few decades the theorem was developed, mainly in Germany, where the mathematical education of engineers seems to have been excellent. Indeed, there is an apocryphal story that a German engineer remarked that continuous beams were popular in countries where the engineers could calculate. The Germans did not think that this group included France, England, or the United States. German mathematician Felix Klein's comment, previously quoted in this text, was not at all apocryphal, however. In a sense there was no cultural imperative for the United States to develop sophisticated theory in structural engineering. Numerous failures, especially in buildings of reinforced concrete, were simply ignored. The age of the pioneers in this material was also a period of numerous collapses.

It is not surprising that the Germans did a great deal of excellent mathematical and theoretical work on indeterminate structures, but the next great step forward came in 1914 from a Dane, Axel Bendixen, using a procedure called "slope deflection." Why did Bendixen use this term? Professor Jack McCormac wrote:

> It comes from the fact that the moments at the ends of the members in indeterminate structures are expressed in terms of the rotations (or slopes) and deflections at the joints. For developing the equations, members are assumed to be of constant section between each pair of supports. Although it is possible to derive expressions for members of varying section, the results are so complex as to be of little practical value. It is further assumed that the joints in a structure may rotate or deflect, but the angles between the members meeting at a joint remain unchanged.[1]

Equations derived from this process express shear conditions for each floor of a building, but it is not very practical for complicated problems. For a six-story building four bays wide, there will be six unknown x values and thirty unknown y values, or a total of thirty-six simultaneous equations to solve. Cross's colleague, Wilbur Wilson, used slope deflection for rigid-frame bridges of reinforced concrete over the Pennsylvania Turnpike. These were simple structures with top and side conceived and built as a single reinforced-concrete unit. The design was so novel that it took many years to gain acceptance, but eventually it became the most popular type of small highway bridge. Cross admitted the utility of Wilson's solution but thought it was inelegant and wasteful of material.

Cross came to the Urbana-Champaign campus of the University of Illinois with an excellent combination of practical engineering experience and academic background. He was already well-known for his lengthy report, *River Flow Phenomena and Hydrography of the Yellow River in China,* and for a monograph used as a text in the Harvard graduate school on the graphical analysis of the elastic arch. With his arrival, the faculty of the civil engineer-

Hardy Cross as a professor at the University of Illinois. Courtesy of the University of Illinois Engineering School.

ing department constituted as gifted a group of men as could be found in that discipline in the United States or Europe. In contrast to many European professors, they all taught. They all published. They exchanged ideas; they criticized one another, and they gave the University of Illinois an outstanding reputation in engineering education. It was no wonder that people came from all over the world to study at Urbana-Champaign.

A chronicle of Cross's years at Illinois notes that he sometimes chose to play the devil's advocate. Once a student named Alford told Cross that he thought one of the problem solutions in their text was wrong. Cross paced back and forth, staring hard at Alford, and pointing at him fiercely. "Can you, a graduate student, actually have the temerity to accuse the internationally known engineer who wrote this book of MAKING A MISTAKE? Can you really believe that the publishers would allow such an alleged error to be printed? Can you show us the error?"

Alford seemed unable to answer.

Still pacing, Cross said, "Can anyone help Mr. Alford? Do any of you see a mistake in problem four?"

The class fell silent.

"Well, Mr. Alford," Cross said sternly, "would you care to retract your accusation?"

"It's just that I can't!"

"Speak up!" Cross thundered.

"I still believe it's wrong!" Alford shouted, his face red with embarrassment.

"Then kindly come to the board and prove it to us," Cross taunted. "We shall be pleased to see the proof of your unfounded allegation."

Alford labored at the board without success for the rest of the period.

Cross began his next lecture by saying, "In our last meeting Mr. Alford raised a serious and unfounded charge against the author of our text." Staring at Alford, he said, "Have you reconsidered your accusation?"

"No sir," Alford replied. "I still believe he is wrong."

"To the board, then, we still await your proof."

Alford's labors were again unsuccessful.

The third time the class met, Cross said, "Mr. Alford, are you ready to withdraw your ill-considered accusation about problem four?"

Moments later Alford was at the board. Within a few minutes he managed to show that the solution to the problem in the book was incorrect, and he returned to his seat. Cross's pleasure was evident from his expression. "You must always have the courage of your convictions," he said. "Mr. Alford does; apparently the rest of you do not, or you are not yet sufficiently well educated to realize that authority—the authority of a reputation or the authority of a printed page—means very little. All of you should hope to someday develop as much insight and persistence as Mr. Alford." Clearly the classroom was an important place for Hardy Cross.[2]

It was also an important place for his great colleagues in the Department of Civil Engineering at the University of Illinois. Of these men, the most important were Arthur Newell Talbot, Wilbur Wilson, and Harald Westergaard. Talbot, who was a native of Illinois, graduated in 1881, spent four years as an engineer with a narrow-gauge railroad in Colorado and returned in 1885 as an assistant professor of engineering and mechanics. The University recognized the importance of his work by establishing a department of Theoretical and Applied Mechanics with Talbot as the chairman. Though not a gifted lecturer, Talbot's accomplishments in research brought him recognition early in his professional career. He developed formulas for relating rates of rainfall runoff to the size of waterways. These formulas became standards for engineering practice. He also developed the "Talbot spiral" or "railroad transition spiral," a method for laying out gradual, jolt-free railroad curves. Talbot achieved international recognition, but says a chronicler " . . . neither he nor any of his contemporaries realized that his greatest claim to fame was the number of outstanding engineers he attracted to his profession and to the tradition of excellence he established at Illinois."

The first of these was Wilbur Wilson, who came in 1913 on a one-year appointment. He performed so well that he soon became a permanent staff member. He shared an interest with Cross in the rigid-frame highway bridge of reinforced concrete, but unlike Cross, Wilson was as interested in the deterioration of bridges as he was in their construction. This interest led him to studies of metal fatigue and ultimately to the design of testing machines, which were used first at Illinois and later in laboratories around the world. Wilson's greatest contribution came in the years leading up to his retirement in 1949. He noticed that in many structures, especially bridges, rivets were failing at an alarming rate. This observation turned his mind to the possibility that high-strength bolts might be more effective as fasteners at the joints. The rivet and bolt manufacturers had faith in Wilson. Their tests agreed with his. Ultimately over four million high-strength bolts were used in the Pan Am building in New York City and three million in the Verrazano Bridge.

The third man to make up the "Big Four" was Harald Westergaard, a Dane who took a Ph.D. in the Copenhagen laboratory of Asgar Ostenfeld. He was working in Germany at the outbreak of the First World War, and as a foreign national he had to leave the country quickly. A friend advised him to go to Illinois and study under Talbot. He did so and was one of the first students to qualify for a Ph.D. in theoretical and applied mechanics at the University. In 1916 he joined the staff. Westergaard was a remarkable man in every way. The historians of the College of Engineering at the University of Illinois note that he was a scholar with an amazing range of interests, an ability to speak and read several languages, and an exceptional talent for theoretical and mathematical problems. Unlike Wilson, Westergaard disliked working with his hands. He had one known affectation: he carried a cane "because one should have an eccentricity to be remembered by." He was as likely to be remembered for his absentmindedness. Sometimes he would drive his car to work, forget it, and walk home.

On the serious side, Westergaard developed an outstanding reputation as an expert in structural mechanics, the analysis of pavement slabs, and the effect of earthquakes on structures. In later years he was chief mathematician for the specialized design problems of Hoover (Boulder) Dam. Part of his work was to determine the degree of earthquake resistance for the intake towers on Lake Mead. Another part involved calculating the stresses that could be imposed on the ground by the world's largest man-made lake.

The thirty-six year old Hardy Cross came into this assembly of notables in 1921. In the notice of his appointment to the Department of Civil Engineering it was observed that he had both practical and academic experience. The University valued both. During the next nine years Cross worked ceaselessly on a method which would enable the structural engineer to analyze statically indeterminate structures more easily than Bendixen's three moment procedure, shown in Appendix C. Cross's breakthrough probably occurred sometime between 1922 and 1924. The best available chronology is given by Cross's colleague Professor Wilbur Wilson in "Some Recent Developments in Methods of Analyzing Statically Indeterminate Structures," a paper delivered at the World Engineering Congress in Tokyo, Japan, in 1929. Professor Wilson stated that he presented the methods described with the permission of Professor Cross. If the chronology offered by Wilson is correct—and there is no reason to think it isn't—three conclusions may be drawn:

* Cross worked tirelessly during the 1920s to refine his moment distribution and column analogy methods in order to present them in 1930 with the maximum effect.
* The graduates of the program in Civil Engineering at the University of Illinois during the 1920s had a considerable advantage over their counterparts in the United States and Western Europe. They knew something that their contemporaries did not.
* Cross wanted the widest possible circulation for his ideas.

Nathan Newmark, who was the most brilliant of Cross's students, believed that the best access to Cross's thought could be obtained by going directly to his published papers. Newmark was certainly correct. This book is based firmly on a study of those papers, which were selected by Newmark and appeared as *Arches, Continuous Frames, Columns, and Conduits* (Champaign, 1963). Cross himself apparently did not want to go to the trouble of writing a textbook, so Newlon D. Morgan, an architect, sat in on his classes and took notes. These notes were the basis of Cross and Morgan, *Continuous Building Frames of Reinforced Concrete* (New York, 1932). The first mimeograph edition of the notes appeared in 1926 and was entitled *Notes on Indeterminate Structures*. It was very popular with engineering students at the University of Illinois—so much so that it was republished in 1950. Professor Emeritus Narbey Khatchaturian, who was part of the republishing project, states that most of the notes were the work of Morgan, not Cross. We would make an exception for the introduction. The writing here is so succinct that it must be the work of Cross. It lays out the nature of his thought in the clearest possible terms. We shall therefore quote from the introduction extensively, briefly consider historical background, and proceed to the papers themselves.

It is evident that Cross believed that the problem of structural indeterminacy was the greatest single issue facing the profession of structural engineering in the 1920s. He must have come to this conclusion on the basis of his seven years of teaching at Brown and his subsequent practice. In an interview of September 13, 2001, Professor Emeritus William J. Hall wisely remarked, "He knew what was out there." No better summary of Cross's effort could be imagined.

Cross first demonstrated the moment distribution method to his graduate students at the University of Illinois in 1922 and the column analogy method in 1924. He first presented them in written form in mimeographed notes for his course in September 1925. Subsequently he embodied his thinking in the previously mentioned *Notes on Statically Indeterminate Structures,* on which he took the trouble to secure a copyright in 1926. The first three pages of those black-bound notes are probably the most succinct account of Cross's thought. We shall deal with them and proceed to the papers as they appeared in Newmark's selection.

In the opening section of his 1926 *Notes,* Cross takes an extremely broad view of his problem in order to emphasize the scope of his work. (Cross was to make his name by applying the theory of elasticity to frames of reinforced concrete, but he was at pains to show that his methods were also applicable to steel.) He divides indeterminate structures into several classes. The most important are:

1. Standard types indeterminate as to reactions and commonly solved as indeterminate structures. These include continuous bridge girders and trusses, including swing bridges and continuous turntables, hingeless and two-hinged arches, and most suspension bridges.
2. More common types in which the external indetermination is often neglected or only approximately allowed for. Under this heading come mill-building bents, elevated-railway bents, portals and similar structures, reinforced concrete building frames, and to a lesser extent steel building frames.
3. Slabs and ribbed slabs in which the main problem is to determine the variations of the shears and bending moments. These include slabs, slabs supported on four sides, and slabs with concentrated loads. An exact solution can be quite complicated, but the laws of statics and a few general principles go far toward providing a satisfactory design.[3]

In the next section Cross takes up the vital issue of what principles are fundamental in the analysis of indeterminacy. He argues convincingly that there can only be one fundamental principle of indeterminate structure and that is *the fact of continuity*. By *continuity* he means the transfer of structural action from one joint of a building frame to another. Until the time of Cross, the fact of continuity had always been considered a major problem for the engineer, especially for one designing in reinforced concrete. He was, of course, thinking primarily in terms of beam theory as it applied to skeleton frames. As we have noted, for the structural engineer the difficulties of calculating such frames in reinforced concrete were formidable. They were, in fact, so formidable that they unquestionably hindered the development of reinforced concrete as a building material. A look at Francis Onderdonk's *The Ferro Concrete Style* (New York, 1928, reprinted Santa Monica, 1998), indicates the paucity of building frames in this material. There is the Franciscan Hotel in Albuquerque, New Mexico (Trost and Trost, 1926). There is a small office building in San Juan, Puerto Rico apparently five stories high, and nine bays long. There is the Dawson Warehouse in Stockton, California, again fairly small. The reason is clear. Unless the moments at the joints of these buildings could be calculated exactly, there was a possibility of structural failure, and no engineer or architect wanted to take that risk. Onderdonk reports that the Hotel Palacio Salvo in Montevideo, Uruguay, was 335 feet high and was supposedly the tallest reinforced concrete building in the world. One wonders what mathematical genius calculated that one. And the scholar will agree with Siegfried Giedion's remark that Alvar Aalto's Paimio Sanitorium of 1929 was "daring." That building was twelve stories high and had a reinforced concrete spine from which the floors were cantilevered outward. The architect's brother was the engineer.

Hardy Cross proposed to change this state of affairs and to show that continuity was a virtue, not a problem. To accomplish this end, he had to cut through a good deal of the conventional wisdom of his day. Part of this thinking was that fiber stresses in a beam could be computed for given moments. If the modulus of elasticity was known, the angle changes resulting from these moments would follow as a matter of geometry. "But," said Cross, "the theory of fiber stress from known moments is not a fundamental of indeterminate structures, and the value of the modulus of elasticity is a fact associated with a definition and not a principle at all, and it is difficult to say what can be considered a fundamental of geometry."[4] Thus Cross disposed of the conventional wisdom on continuity.

In the next section of his introduction Cross takes up the complex question of the nature of theory in engineering. His words are still worth pondering:

> It is perhaps worthwhile to call attention to the double use of the word "theory" in scientific discussions. In some cases it is used to mean a body or group of facts the truth of which is not questioned; in others it means a hypothesis which has strong evidence in its favor though its truth is still open to some questions. Thus the theory of elasticity is a group of geometrical relations which are not open to debate, but the idea that time yield of the concrete will delay failure from temperature stresses in a concrete arch is a theory in quite a different sense. Other debatable points in indeterminate structures are not theories at all, but merely convenient assumptions; thus no one holds any theory that the modulus of elasticity is constant throughout an arch ring, the only question being whether such variations as do occur produce any important effect on the results.
>
>
>
> Much confusion of thought has come from misuse of this term. We may further cite: as groups of facts not open to experimentation or debate, the theory of the elastic arch, the theory of continuous girders, the theory of deflection; as hypotheses strongly supported but as yet not fully proved, the theories of fatigue failure, the theory of earth pressure, the theory that the strength of concrete in a structure is the same as that shown by a cylinder in a testing machine or that rate of application of load is a negligible factor in producing failure; and finally, as misuses of the word, the "theory" that the moment of inertia of a concrete beam varies as bd3, that the tension rods do not slip in concrete beams, that there is no distortion due to shear. The first group of "theories" is not debatable, the second depend usually on experimental verification, while in the case of the third the important question is, how significant is the error? The data often needed in the third group are elementary; when these are available, deductive processes furnish a definite answer as to the importance of the error.[5]

Though not formally trained as a physicist, Cross may be said to have been primarily concerned with the application of the theory of elasticity to structural design. The theory of elasticity is a branch of physics which goes all the way back to the *Dialogues Concerning Two New Sciences* of Galileo (1638). Various physicists and engineers have contributed to it since that date. The contributions of Newton and Hooke were tremendous. For Cross's purposes, the most important commentator was probably James Clerk Maxwell who, in 1864, noted that: *The deflection of one point A in a structure due to a load applied at another point B is exactly the same as the deflection at B if the same load is applied at A.* This rule is perfectly general. It applies to any type of structure, whether it is truss, beam, or frame, which is made up of elastic materials following Hooke's law. Maxwell's observation introduced the first consistent method for analyzing indeterminate structures, but his presentation was brief and did not attract much attention. Ten years later Otto Mohr independently extended the theory almost to its present stage of development.

Cross also considered objections to his proposed application of the theory of elasticity to structural design. He admitted its difficulties but believed that his methods generally overcome them. He says:

> Reluctance to apply the theory of elasticity to structural design is based on two objections, one that it is too arduous to be justified by its results, the other that it is not sufficiently flexible to permit evaluation of the effect of such factors as variable E, brackets, gusset plates, imperfect elasticity and phenomena beyond the yield point. If the latter objection holds, the former becomes important since we have only our labor for our pains. In many cases the objection is valid, and this tool of design is useful only in proportion as it permits the evaluation of the effect of physical uncertainties. By the methods presented in this text, such uncertainties can be included in the analysis of plane continuous frames. This is a fruitful field of research. In some cases large uncertainties in the data may be shown to produce small variations in the results. In other problems, notably in the case of deformation constants in arch foundations, the uncertainties may so seriously affect the results as to make precise computations illusory.[6]

Cross admits that in many cases the objections are valid, but he believes that his methods can overcome physical uncertainties. He based himself firmly on geometry and put his case in the following language:

> The study of statically indeterminate structures will have at its foundation two elementary fundamental conceptions.
>
> 1. Any point at rest is in static equilibrium.
>
> 2. Any line that is continuous preserves its elastic properties.
>
> From these two simple basic ideas follow the laws of statics and the laws of continuity. Statics gives us the three conditions of equilibrium of nonconcurrent forces: $H = 0$, $V = 0$ and $M = 0$. From continuity follow simple geometric relations between the displacements and angle changes in the structure.
>
> It should be emphasized that the principles of geometry and the most elementary calculus together with the laws of statics are sufficient for solving all the ordinary problems in indeterminate structures that occur in practice. It is of vital importance in the solution of such problems to be able to visualize the action of the structure under load. A *qualitative* study of the probable action of the structure will invariably lead to the quickest estimate of the *quantities* involved. A qualitative sketching of influence lines and of the approximate shape of the deformed structure will not only make it possible to avoid certain

unnecessary lines of investigation as irrelevant to the problem in land but will in many cases reveal the critical point or points to be investigated and usually the critical loading involved.

Certainly the first thing and perhaps the most important thing to be considered in the design of any structure is the determination of what needs to be figured—what is worthwhile. Associated with this question is that of the precision that is desirable as well as that which it may be practicable to obtain.[7]

Here Cross is simply restating Newton's principles. He also shows himself to be a firm believer in graphic statics, a discipline first developed by Professor Karl Culmann at the Eidgenossische Technische Hochschule, in Zurich, Switzerland.[8] It is no wonder that Robert Goodpasture recalls that "Cross was always a great one for drawing pictures." With a properly drawn picture an overly complicated mathematical analysis can often be avoided.

Thomas Golden Jr., another of Cross's students at Yale, recalls:

[Cross] was given to explaining by the simplest of means such as taking and bending a wire coat hanger to portray the action of a continuous beam. His moment distribution and column analogy were really shortcut math techniques to solve analytical problems previously approached with cumbersome and probably no more accurate solutions to the same problem. And his moment distribution had application to the action of fluids in closed systems as well and perhaps other things . . .

I do know Cross was very much taken with the early development of reinforced concrete and used to dwell on the intellectual ironies of employing sophisticated mathematical techniques to analyze the action of continuous concrete members when the underlying strength of the components was only within a range of prediction.[9]

Cross himself was primarily an analyst but clearly felt that there was not enough good design in the engineering world. He dealt with that problem forcefully:

Limitations—Physical and Mathematical. The designer will need to keep clearly in mind two entirely independent types of limitations:

1. Limitations of a physical nature.
2. Limitations of a mathematical nature.

These two are sometimes confused with annoying consequences. It is worthwhile at all times to realize the fact that such things as moment, shear, stress, moment of inertia, etc., are merely definitions—convenient mathematical expressions, not physical phenomena. Loads are not transmitted nor are stresses carried by this member or that, and while such expressions may be justified by

The Creative Years at Illinois

their convenience, it is important that our mental picture be accurate and that we do not lose sight of the actual geometry of the problem.

Physical limitations such as the action of rivets and the effect of gusset plates in steel design, the effect of spandrel columns, of T-slabs, time yield and variations of E. in concrete, all such, need to be distinguished in our thinking from limitations of a purely mathematical nature. Each will affect the significance of the numerical result but in a different way.[10]

Newmark points out that, Cross dealt with the problem of structural design as contrasted with analysis in three important papers. He stated his own method in the following provocative language:

General Method of Analysis. The common method of attacking problems in indeterminate structures has been based on the general idea of making the structure statically determinate by cutting redundant members at points of continuity, and then tying these static groups together by the principles of continuity, solving these resulting groups of equations simultaneously for the various unknowns. The method outlined in this text is usually the reverse: preserving at all times the continuity, distributing the unbalanced moments according to the laws of that continuity, and finally adjusting as need be, to conform to the laws of statics.[11]

Evidently Cross disregarded the conventional system of analysis and offered something entirely new. It was this novelty of approach which caught the imagination of the engineering world. Cross threw away all kinds of customary procedures and offered a system which was both simple and useful. It was also more elegant.

Finally Cross stated:

Three Important Principles. The method here employed of attacking indeterminate structures involves three easily established principles which may be summarized as follows:

1. Column Analogy. If any restrained elastic ring (fixed beam, bent, arch, etc.) is treated in its outline as the cross-section of an analogous column whose differential areas are the ds/EI values of the ring, the bending moments due to continuity at any point are analogous to the corresponding fiber stresses in the column due to the angle changes as loads.

2. Distribution of Moments and other forces. If the fixed end moments are computed for any joint the unbalanced moments may be distributed among the bars in proportion to their stiffness, and a portion carried over to the other end. This involves the distribution of a succession of unbalanced moments, but the method is quite general in its application and the series commonly converge rapidly to any degree of accuracy desired. In

its general application it will be seen to include, with slight modification, the distribution of joint rotations, displacements and shears, as well as the distribution of moments.

3. Virtual work. Displacement, linear or angular, at any point in a loaded structure is equal to the virtual internal work—this being described as the sum of the products of the existing distortions and the virtual resistances to a unit coincident force of displacement. The reactions to this hypothetical force of displacement will fix the reference by which the displacement may be measured. The direct application of these principles coupled with the laws of statics affords an immediate solution of most of the ordinary problems in indeterminate structures. Variable moment of inertia offers no peculiar difficulty. Signs may be readily taken care of by the usual conventions. Indeed in many cases they may be determined by inspection at the end of the solution. The most common language of the engineer—moments, shears, stresses—constitutes the greater part of the work. The relation of the structure is readily visualized and limitations, mathematical and physical, become apparent. Close approximate results may be obtained by the same processes which if carried further will give any degree of precision.[12]

Each one of these principles was a topic in the lecture notes which followed Cross's introduction, and each became the subject of one of his three major papers.

The statement of Cross's three principles in the previous paragraphs is an outstanding example of his clarity and precision in technical writing. In the publication of his paper, "Virtual Work: A Restatement" in the A.S.C.E. *Proceedings,* February, 1926, he goes into the problem of virtual work at greater length, and credits Professor George F. Swain with the first presentation of the principle in American engineering literature in February, March, and April 1883. So it seems reasonable for Cross to state, "Most of the theorems developed are not new. Indeed, the writer is not sure that any of them are entirely new, nor is he here concerned with their lack of novelty."[13] In his approach to the problem he somewhat resembled a composer interested in counterpoint who goes back to see what Bach had to say on the subject.

In the body of the paper, Cross first gives the usual statement of virtual work in terms of the conservation of energy, then gives his restatement, and proceeds to a consideration of virtual work and to a treatment of beams and trusses. In the light of his restatement he takes up Maxwell's theorem of reciprocity and Mohr's improvement. He also includes provocative sections on angle changes in trusses and slope differentiation, and concludes with the crucial observation that virtual work is an excellent method for computing moment weights in trusses. He offers an excellent illustration as well:

Assume that it is desired to draw the deflected load line for a unit load as shown (fig. 1), that is, an influence line for horizontal reaction. The angle change may be computed by applying a unit moment resisting this angle change. This is effected by applying loads as shown in Figure 1 (b), acting at the points on the load line or floor as indicated by circles, and then computing $\Sigma\delta v$, in which, Σ is the change in length of any bar due to the horizontal reaction, and v is the stress for the loading shown in Figure 1 (b). These angle changes may then be treated as loads at the panel points and, when corrected for the deflection of the ends of the load line, the moment curve thus produced will have the shape of the influence line.[14]

We have here a beautiful example of Cross's adherence to the method of graphic statics.

In order to develop his argument on moment distribution, Cross needed an additional tool which would simplify the finding of fixed-end moments. He found it in "The Column Analogy," the second in his series of essential principles listed in his 1926 publication. Most of Cross's great papers are brief, but the paper on "The Column Analogy" is an exception. It required seventy-eight pages in the Newmark collection and Cross himself called it a monograph. The author published his findings in *Bulletin 215* (1930) of the Illinois Experiment Station. It showed that " . . . bending moments in arches, haunched beams, and framed bents may be computed by a procedure analogous to the computation of fiber stresses in short columns subject to bending, and that slopes and deflections may be computed as shears and bending moments, respectively, on longitudinal sections through such columns."[15] He proceeded to take a sensible view of the analogy of the moments produced in an indeterminate structure and the stresses produced in an eccentrically loaded short column. His method was applicable to structures indeterminate to not more than the third degree. These included fixed-end beams, single span arches and frames, and closed boxes. Equally valuable was another application of the column analogy: the calculation of the carry-over factors, stiffness factors, and fixed-end moments necessary

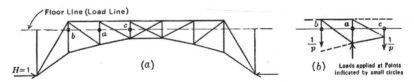

Figure 1. Diagram showing computation of moment weights. From Hardy Cross, *Arches, Continuous Frames, and Conduits*. Introduction by Nathan M. Newmark. University of Illinois Press, Urbana, 1963.

for analyzing structures with members of varying moments of inertia by moment distribution. For an application of column analogy, see the illustrations provided by Professor Kemp in Appendix C.

Again Cross argued that the problem was basically a matter of geometry. "If the elastic properties of the different portions of the structure are definitely known," he wrote, "the analysis of restrained members is essentially a problem in geometry, because the member must bend in such a way as to satisfy the conditions of restraint. The geometrical relations involved are identical in algebraic form with the general formula for determining the fiber stress in a member which is bent."[16] Thus the analogy pertained to the identities existing between the moments produced in an indeterminate structure and the stresses produced in an eccentrically loaded short column. Every capable engineer knew how to handle the latter problem. Once Cross had pointed out the identities, the computations became easy. He seems to have always worked with the engineer-in-practice uppermost in his mind. Column Analogy is taught even today in several schools of engineering because it is one of the simplest tools for developing stiffness coefficients for certain applications in finite element analysis. Cross was closing in on his greatest contribution, his method of moment distribution.

So far in this chapter there have been only a few references to the materials with which Cross was dealing: timber, steel, and reinforced concrete. It was the rapid development of this last material in the early years of the twentieth century that made a simple method for calculating its behavior a necessity. The late Reyner Banham was certainly correct in stating that around 1910 the leading architectural and engineering firms in the United States switched to concrete framing for industrial buildings. His favorites among these offices were those of Ernest L. Ransome, Albert Kahn, and Lockwood Greene, and Co. of Boston. Banham did a brilliant job of analyzing the innovations brought about by these pioneers. In their work it became clear that reinforced concrete was a material whose potentialities were almost unlimited.[17]

It was one problem, however, to build a daylight factory of two or three stories with a reinforced concrete frame and quite another to use such a frame in a bridge, an office building, or an apartment block. In these structures the engineering requirements seemed to dictate that increasing loads would be met by increasing the size of the structural elements. In the period from 1910 to 1930, wrote Carl Condit, "the result, which was on an elephantine scale, especially in buildings with large, open interiors, revealed the diversity of requirements that could be satisfied by reinforced concrete but also the amount of valuable space that was consumed by the sheer quantity of material in long beams and trusses."[18] Engineers and architects were properly very conscious of the safety factor. Hence there was a tendency to over-design the structural elements and create buildings which were very heavy "on an elephantine scale." With his usual acuity, Hardy Cross noted that concrete was much heavier in comparison with its strength than timber or steel. "Its weight," he said, "eats up its strength in long span construction."[19] Furthermore, it looked clumsy unless designed with skill. Essentially Cross's grand objectives were to overcome this state of affairs and to simplify the design of statically indeterminate structures.

There was, therefore, an intense interest in reinforced concrete among architectural designers, especially of the modernist persuasion, in the United States and Europe during the period from 1910 to 1930. The key book in defining the new architecture which had emerged after the First World War was unquestionably *The International Style* by Henry-Russell Hitchcock and Philip Johnson. It appeared in conjunction with an exhibition at the new Museum of Modern Art in New York in 1932. While the principal concern of the authors was, as Alfred Barr remarked, "aesthetic quality," they could not entirely escape some consideration of structure. In one important passage they pointed out that with the use of steel or reinforced concrete skeletons, modern builders could equal the most daring feats of Gothic architects. Insofar as one can judge by the examples which were offered, reinforced concrete was, in fact, the canonical material for International Style practitioners. Brick was passé. Timber was simply out of the question. From the photographs in the Hitchcock-Johnson opus, we would say that most of the major buildings were done in column and slab concrete. The largest single building in the volume was probably the Van Nelle Factory at Rotterdam (Van der Vlught and Brinkmann, 1930). Photographs clearly reveal that, for this eight-story building, the concrete flat slab was the primary structural device (Figure 2). Professor Niels Prak thought that the system was "American." Indeed the capitals do bear a distinct resemblance to the familiar mushroom forms of Claude A. P. Turner, but they might also derive from Maillart in Switzerland. One might note in passing that the famous Mies van Der Rohe project for a concrete office building of 1922 made a similar use of the column and slab. Even the well known Bauhaus at Dessau (Gropius and Meyer, 1926) was done in this manner. None had what we would today recognize as a continuous rigid frame.

The reason for this reliance on the column and slab is not difficult to understand. In the 1920s the structural behavior of the column and slab was better understood than that of the continuous rigid frame. After Westergaard and Slater's classic paper of 1922, the true nature of the column and slab was clear to every structural engineer on both sides of the Atlantic.[20] But the difficulties with the column and slab were that it could not be used for

Figure 2. Van Nelle Factory, Rotterdam by Van der Vlught and Brinkman, 1929. Nederlands Architectuur Instituut, Rotterdam.

wide span solutions, it required a great deal of material, and unless designed with the artistry of a Robert Maillart, it was a clumsy system.

Obviously structures with continuous rigid frames were desirable aesthetically and economically, but the problems of engineering analysis and design were formidable. Cross discussed their nature in "Why Continuous Frames?" American Concrete Institute, *Proceedings,* 31 (1935). He illustrated his argument with four drawings (Figures 3–6). The first, Cross said, might be called the post and lintel stage of reinforced concrete. The second was a kind of hybrid—a rigid frame, but a very poor rigid frame. It bears a marked resemblance to the kind of small highway bridge developed by Cross's colleague, W. B. Wilson. The third is a similar type of rigid frame building. Cross believed that a three-hinged arch was the most efficient form. This closely resembles the structure of the famous Galerie des Machines (Cottancin and Dutert, Paris, 1889). Finally he showed a beam with constant section and a beam with a parabolic soffit tapering to a very small depth at the center. This last has a curious resemblance to the kind of form employed by Maillart in several of his best-known bridges (Figure 7). Cross noted that the amount of material required in the second case is only $1/12$ of that in the first case and the amount of reinforcement is $1/4$ as much. He readily admit-

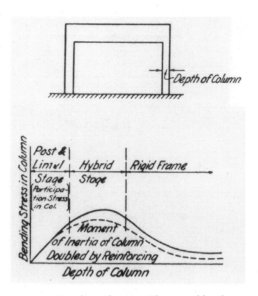

Figure 3. Two-legged Bent with vertical loads. The girder has been designed as a beam simply supported. It is assumed that the columns width is constant. Reinforcement is neglected. Cross, *Arches, Continuous Frames, and Conduits.*

Figure 4. Rigid frame with column depth increased. Hybrid stage of action. Cross, *Arches, Continuous Frames, and Conduits.*

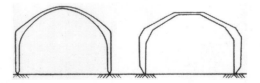

Figure 5. Rigid frame building with three-hinged arch. Cross, *Arches, Continuous Frames, and Conduits.*

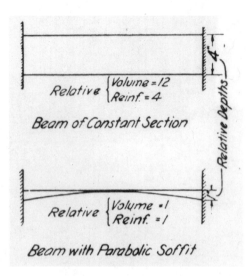

Figure 6. Beam of constant section and beam with parabolic support. Cross, *Arches, Continuous Frames, and Conduits.*

Figure 7. Salginatobel Bridge, Maillart, 1937, Grant Hildebrand.

ted that his conclusion was not to be taken too seriously because no live load was involved. But it did "challenge attention." My own feeling is that Cross was fascinated by Maillart's forms but never called on him in Switzerland.

One of the great mysteries in Hardy Cross's career is his apparent unwillingness to comment on Maillart's bridges or on any of the other masterworks in concrete by Torroja, Nervi, and Candela. The first notice of Maillart in an American engineering periodical was an article by M. S. Ketchum Jr., "Thin-Section Concrete Arches as Built in Switzerland," *Engineering News Record,* Jan. 11, 1934, 44–45. The article contained photos of both the Klosters and the Valtschielbach bridges, and was in a periodical customarily read by Cross. We know that Professor and Mrs. Cross traveled extensively in Europe during the 1930s, and that Cross was a connoisseur of bridges. Did they go to Switzerland? We do not know, nor do we know if they were in those parts of Italy where they might have seen the early works of Nervi.

The essential structural fact Cross confronted was that monolithic concrete frames are highly indeterminate. In order to build them there was a need for methods of analysis which gave reasonably accurate results and did not require a horrendous amount of calculation.

The slope deflection method developed by Axel Bendixen in 1914 (see Illustration in Appendix C) was the first really practical way to solve rigid-frame structures. His method leads to an easily written series of simultaneous equations. Each equation is rather "sparse" (i.e. it contains only a few of the unknowns). Thus the effort required to solve these simultaneous equations was less than methods developed earlier. The results from the solution of the simultaneous equations yielded rotations and displacements at the ends of individual members that in turn could be used to find moments and shears, but this procedure prior to the advent of computers was laborious and time consuming.

In 1922 K. A. Calisev, writing in Croatian, offered a method of solving the slope deflection equations by successive approximations. Hardy Cross seems not to have been aware of Calisev's contribution, probably because of the linguistic difficulty. Though cumbersome, it was a pioneering work. The problem with it was that Calisev still used successively adjusted rotations to establish moment balances at the nodes.

Although Cross was known as philosophical, his approach here was extremely practical. His view was that engineers lived in a real world with real problems and that it was their job to come up with answers to questions in design even if approximations were involved. In his first paragraph he wrote, "The essential idea which the writer wishes to present involves no mathematical relations except the simplest arithmetic."[21]

The Cross method depended on the solution of three problems for beam constants: the determination of fixed-end moments, of the stiffness at each end of a beam, and of the carry-over factor at each end for every member of the frame under construction. The fixed end moments, stiffness, and carry over factors can be easily found from the slope deflection equations, area moment theorems, etc., without using calculus, a branch of mathematics in which too many engineers were weak. The engineer could use standard formulas to carry out computations employing arithmetic. Its application required no contrivance more complicated than a slide rule.

Perhaps because of his background in the humanities, Cross's language in his technical papers was always remarkably succinct. In Cross's later years a visitor to his offices at Yale found *The Theory of Elasticity* standing next to Carl Sandburg's *Abe Lincoln in Illinois, The Holy Bible,* and *Alice in Wonderland* on his library shelf. The last three titles are certainly notable for their stylistic qualities, as much engineering literature, alas, is not. Thus, for anyone who knows what a moment is, the description of Cross's moment distribution method is a little masterpiece.

It was Hardy Cross's genius that he recognized he could bypass adjusting rotations to get to the moment balance at each and every node. He found that he could accomplish the same task by distributing the unbalanced moments while unlocking one joint at a time and keeping all the others temporarily fixed. By going around from joint to joint, the method converged very fast (at least in most cases) and it had enormous psychological advantages. Many practicing engineers had dubious mathematical skills in handling simultaneous equations, and many had difficulties in visualizing rotations and displacements. Moments are much more "friendly" for the average engineer and therefore easier to deal with. Thus the Moment Distribution Method (also known as the Cross Method) became the preferred calculation technique for reinforced concrete structures. He wrote:

> The method of moment distribution is this: (a) imagine all joints in the structure held together so that they cannot rotate and compute the moments at the ends of the members for this condition; (b) at each joint distribute the unbalanced fixed-end moment among the connecting members in proportion to the constant for each member defined as "stiffness"; (c) multiply the moment distributed to each member at a joint by the carry-over factor at the end of the member and set this product at the other end of the member; (d) distribute these moments just "carried over"; (e) repeat the process until the moments to be carried over are small enough to be neglected; and (f) add all moments—fixed-end moments, distributed moments, moments carried over—at each end of each member to obtain the true moment at the end.[22]

In the next paragraph Cross observed that for the mathematically inclined his method would appear as a means of solving a series of normal simultaneous equations by successive approximation. Indeed it was. His willingness to accept approximation troubled that minority in the engineering profession who insisted on exactitude. Cross himself readily admitted that the values of moments and shears could not be found exactly. He concluded that there was no point, then, in trying to find them exactly.

The greatest contribution of Hardy Cross has an amusing background. His dean, Milo Ketchum, resented Cross's reputation as a great teacher. During his term of office Ketchum prevented Cross from getting salary increases and several times suggested that the professor might do better working elsewhere. One of Ketchum's charges against Cross was that he failed to publish enough papers, although in 1926 he had already published his noted paper, "Virtual Work: A Restatement." In any event, Cross's response to Ketchum's threats was the publication of a ten-page paper in the May 1930 *Proceedings* of the American Society of Civil Engineers entitled: "Analysis of Continuous Frames by Distributing Fixed-End Moments." It set forth an entirely new method of analyzing building frames. Discussion was closed in September 1930. The paper was again published in the *Transactions* in September 1932. At that time, space was afforded for 38 commentators who took up 146 pages. This may be a record for comment on a single paper. Cross was immediately hailed as a man who had solved one of the knottiest problems in structural analysis, and he did so in a way that could be adopted by any engineer working in the field.

Cross was after a method of analysis which combined reasonable precision with speed. That he achieved his end is clear from a survey of the thirty-eight commentators to his brief article. These individuals, some academics, some practicing engineers, generally saw that his method was a major contribution to structural analysis. It quickly passed into the curricula of the best American schools of engineering and by 1935 was commonly taught. It continued to be received doctrine until the 1960s. Cross did all this with just two diagrams and one table (Figures 8, 9, 10). The first diagram is generalized to the highest possible degree. Its only requirement is that the members be straight and of uniform stiffness. The table speaks for itself. The second diagram shows a possible correction for side-sway.

In a certain sense Cross gathered ideas which had been circulating for several years. There was nothing new about the notion of distribution. There was nothing new about a mathematical series which was convergent. Cross simply put the two together. Important scientific discoveries often have this character. For Einstein's theory of relativity the experiments of Michelson

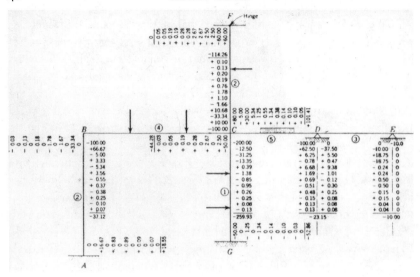

Figure 8. Diagram for moment distribution. Cross, *Arches, Continuous Frames, and Conduits.*

ccessive values of bending oment at joint.		After one distribution (two rows of figures).	After two distributions (four rows of figures).	After three distributions (six rows of figures).	After four distributions (eight rows of figures).	After five distributions (ten rows of figures).	After six distributions (twelve rows of figures).
A		0	+ 16.67	+ 17.50	+ 18.39	+ 18.48	+ 18.55
B		− 33.34	− 35.01	− 36.79	− 36.97	− 37.10	− 37.13
C	In CB	− 90.00	−112.66	−113.22	−114.24	−114.23	−114.26
	" CF	+ 75.00	+ 99.66	+100.36	+101.82	+101.36	+101.41
	" CD	−212.50	−257.10	−258.09	−259.89	−259.88	−259.93
	" CG	− 47.50	− 44.83	− 44.55	− 44.36	− 44.31	− 44.28
D		− 37.50	− 32.03	− 23.66	− 23.48	− 23.15	− 23.15
E		− 10.00	− 10.00	− 10.00	− 10.00	− 10.00	− 10.00
F		0	0	0	0	0	0
G		− 50.00	− 51.25	− 52.59	− 52.73	− 52.83	− 52.86

Figure 9. Convergence of results. Cross, *Arches, Continuous Frames, and Conduits.*

and Morley were important. So were the mathematics of Minkowski. Genius seems to consist in seeing the relation of ideas. It was crucial that Cross had this ability. He also stated his case in such fashion as to invite shortcuts and emendations. In a reply to a certain Professor Pilkey, who complained that the paper was too short, he quoted Pascal's famous remark in *The Provincial Letters* (1656). "I have only made this letter rather long because I had not the time to make it short." Cross added that he had taken the time to make the paper short and, as he looked at the mass of the technical literature on his desk, he could only hope that this idea would become popular.

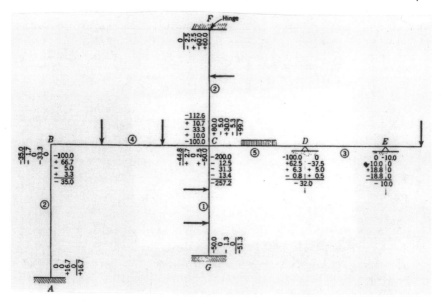

Figure 10. Correction for sidesway. Cross, *Arches, Continuous Frames, and Conduits*.

Another writer suggested that the method might have important applications in aircraft design. This was a prophetic insight. In October 1937, Eugene E. Lundquist of the Langley Aeronautical Laboratory used the Cross method in "Stability of Structural Members Under Axial Load." It is no wonder that the professor is said to have remarked with the publication of his paper that he believed his faculty could bear this particular Cross.

Although the Cross method was primarily helpful to designers working with frames of reinforced concrete, it was also useful to those creating buildings and bridges of steel. It was only necessary that the frame be rigid, and this quality could be secured with the new bolts of high-strength steel devised by Wilbur Wilson. These were used in the Verrazano Bridge. An anecdote is told of a conversation between Mies van Der Rohe and Frank Kornacker, his engineer for the Seagram Building (New York, 1956). Because of wind loads, the New York engineers wanted to weight the whole building, put shear walls in, and load it with concrete so that the mass would resist the wind forces. "Mies," said Joe Fujikawa, "thought that was awful and a primitive way of doing a structural design. So he had Frank Kornacker do an analysis. Frank thought a better structure would be to make a moment frame out of it and use rigid column and girder connections to pick up the moments and not limit lateral resistance to shear walls at the elevators." Such was the origin of the much-admired plan of the Seagram Building. Mies van Der Rohe firmly believed in the expression of structure. It could easily be argued that in the Seagram Building he was expressing the moment frame developed by Hardy Cross.[23]

There remain two curious questions about Hardy Cross's moment distribution method. In 1949 there was a symposium at Illinois Institute of Technology dedicated to Cross. It was convened by Lawrence P. Grinter, who also edited the papers which were presented under the title *Numerical Methods of Analysis in Engineering* (New York, 1950). The volume was dedicated to Hardy Cross, who gave us " . . . a simple demonstration of the power of numerical analysis." This is very strange language in a book honoring a man who was all his life an advocate of the power of graphic statics. Among those present, however, was Richard Southwell from England. His relaxation method was similar to Cross's.

The second question concerns the treatment of Cross by Stefan Timoshenko in his monumental and authoritative *History of Strength of Materials*. It is limited to one short paragraph on page 424. It disposes of the Cross contribution in a few perfunctory sentences. Timoshenko, whose prestige was immense, remarks that " . . . quite a bit of building was done in the United States according to this method."[24] It is almost as if Timoshenko had some kind of grudge against Cross. We will never know because both men have passed away.

Born in Russia in 1878, Stefan Timoshenko was the son of a surveyor. In his childhood he showed an interest in mathematics and solved various problems not because they were assigned but because he enjoyed them. At the age of fourteen he learned how to sketch and draw, and he participated in the building of a house. During the summers of 1899 and 1900 Timoshenko went to work on the Volchursk-Kupyansk Railroad in order to learn the practical aspects of construction.

Upon graduation from the gymnasium, he did his military service and at its completion married Alexandra Archangliska, a medical student. For a time he worked at a mechanics laboratory where he became familiar with testing machines. However, he soon became convinced that for scientific work a more thorough grounding in mathematics was needed. In 1904 he went to Germany with the specific purpose of becoming better acquainted with German technical schools and their teaching methods. At this time he was also interested in Lord Rayleigh's book on *The Theory of Sound* and was particularly captivated by Rayleigh's approximate methods of calculating vibration frequencies in complex structures. In the 1907–8 school year he gave his first full course on the strength of materials. The outbreak of the

The Creative Years at Illinois

First World War found him installed as a Professor of Communications and Electrical Engineering at the Petersburg Technological Institute.

Out of sympathy with the Revolution (his brother became a Soviet field marshal), Timoshenko left Russia to become Professor at the Zagreb Polytech. In a few years he moved to the United States, where, for a time, he worked for Westinghouse. In the United States, Timoshenko felt that the thoroughness of his training in mathematics and basic engineering subjects gave him a tremendous advantage over Americans, especially in solving non-stereotypical problems.

In 1927 he accepted a special chair of Research in Mechanics at the University of Michigan. At Ann Arbor his lectures on Applied Mechanics became famous. He attracted a large number of students from other departments and also young teachers. A summer session was instituted, and a number of distinguished engineers from other institutions were attracted. This included Prandtl from Göttingen, Southwell from London, Von Karman from Cal Tech, and Westergaard from the University of Illinois. During his tenure at Michigan, Timoshenko published books on the strength of materials, the theory of elasticity, and elastic stability.

Attracted by a salary increase and by the California climate, Timoshenko took a position at Stanford in 1936. There he published books on engineering mechanics, the theory of plates and shells, and advanced dynamics. His last book was a *History of Strength of Materials*. It traced the development of his discipline from Leonardo da Vinci and Galileo to the present. In the course of his career he was elected to many learned societies such as the National Academy of Science of the United States and the Royal Society in London. He received honorary degrees from various universities such as Lehigh, the Swiss Technological University in Zurich, and Glasgow University. In 1935 the American Society of Mechanical Engineers conferred upon him its Worcester Reed Warner Medal for achievement in the field of mechanics. Stefan Timoshenko died in 1954.

A more discriminating evaluation of Cross's work, one which is probably typical, is to be found in Jack C. McCormac's *Structural Analysis* (Scranton, 1962). The author writes that the Cross paper, "Analysis of Continuous Frames by Distributing Fixed-End Moments" started " . . . a new era in the analysis of indeterminate frames and gave added impetus to their use. The moment distribution method of analyzing beams and frames involves little more labor than the approximate methods but yields accuracy equivalent to that obtained from the infinitely more laborious exact methods previously studied."[25] As one might expect in a textbook, McCormac's explanation is considerably more detailed than that of Cross, and includes a picture of a steel-framed office building at 99 Park Avenue in New York City. The Cross method was a blessing not only to engineers and architects working in reinforced concrete but also to those concerned with steel. The only requirement was that the frame be rigid and continuous. McCormac also included a chapter on the column analogy. He saw it as the most convenient method of calculating carry-over factors, stiffness factors, and fixed-end moments.

The moment distribution method began to lose its enormous popularity when computer programs became available. However, the method is still used for preliminary analysis of simple indeterminate beams. Secondly, the basic solution by the stiffness method that is now commonly the basis of matrix-based computer analysis begins with the step that the structure is initially locked, i.e., all joints are fixed against displacements and rotations; stiffness coefficients are then calculated by displacing each fixed joint, one at a time, through serial displacements and finally equations of joint equilibrium are written and solved to find the joint displacements. This idea of locked structure owes its origin to Hardy Cross, whose moment distribution method also has a first step where the structure is initially locked at the joints.

The last paper that Newmark selects has a curious history. It was written quickly at the request of a few colleagues and is entitled, "Analysis of Flow in Networks of Conduits or Conductors," and appeared in the University of Illinois experiment station *Bulletin* in 1936. At the outset, Cross noted that the problem of finding the distribution of flow in networks of pipes is a design problem in systems of water distribution. He added that similar problems occur with the distribution of steam or air and with electrical circuits. The analysis of such systems by formal algebraic methods was difficult if the networks were complicated. There was some value in models, but there were objections to their use. Once again Cross offered methods of successive connections and convergence.

He believed that the convergence was sufficiently rapid to make the procedure useful in practice. And as always, he held that the greatest value of his analytical method was in training and assisting judgment. The problem to which Cross addressed himself may be stated thus: in a water supply system, how can one determine the loss of head that will occur through any looped network of different size and type of pipe? To solve this, Cross turned to a combination of geometry and the differential calculus. Essentially he developed a clever use of the binomial theorem to secure an iterative process which would converge quickly if the proper assumptions were made. Today it is taught at the U.S. Air Force Academy where it is known as the Hardy-Cross [sic] method for balancing flows in distribution systems.[26] The procedure is also taught at the University of Massachusetts, Auburn, the University of

Central Florida, and Queen's University. Various types of software illustrating the method are for sale on the Internet. It does not appear that this contribution of Hardy Cross's to hydraulic engineering will go out of style very soon (Figure 11). In most texts on structural analysis Hardy Cross's moment-distribution method will be mentioned with enormous respect (if the authors have any sense of history), but there is no doubt that the computer has provided much faster and more powerful methods of analysis. Forty-two years after his death, however, the Hardy Cross method for the computation of flow in hydraulic networks is very much a part of American engineering practice. We know of no similar episode in the history of the discipline of civil engineering.

After the appearance of his "Moment Distribution Paper," Cross received many invitations to speak before engineering societies and to write for various publications. Perhaps his most provocative essay was a remarkable summary of "Structural Knowledge." The article dealt with developments in the preceding year and it appeared in *The Engineering News Record* February 6, 1936 (199–200). It is worth quoting in its entirety since it shows his keen awareness of everything that was going on across the country.

> In the field of structural engineering the past year has been one of consistent development rather than of spectacular accomplishment. With its usual vigorous digestion, the structural art has consumed masses of analytical material

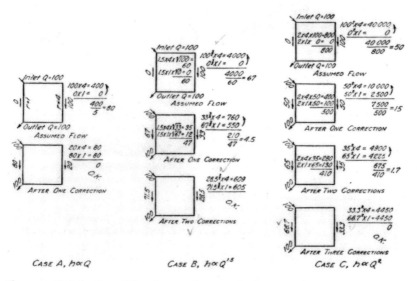

Figure 11. Distribution of flow in sample network. Cross, *Arches, Continuous Frames, and Conduits.*

and results from model studies, data from laboratory experiments on materials and on details. The results of this digestive process will appear in structures that are yet to be built.

Most important completed bridges of the year are the great New Orleans crossing of the Mississippi—a structural dream for over 50 years—and the suspension bridge at Davenport. At our two great ports, the Triborough Bridge is under construction at New York, Golden Gate and the great East Bay cantilever at San Francisco. The giant builders are still with us; to realize this, compare the moderate publicity given to the 500 ft. continuous spans at South Omaha with that which they would have received twenty years ago!

NOTABLE RESEARCH WORK

The engineers of the San Francisco-Oakland Bay Bridge praiseworthily recognized their indebtedness to the engineers of the past by financing extensive tests of riveted joints. Large joints under static load are being tested at the University of California; fatigue tests of smaller joints are in progress at the University of Illinois. These promise to rank among the greatest experimental researches in structures. Most important, however, is recognition that each major structural project has a debt to the past that can be paid only by a contribution to the future.

Reconsideration of fatigue as an element in design is shown also by tests on reversed loading of welded joints at the University of Illinois and by renewed consideration of the subject by technical committees. Some will recall discussions of the topic at the beginning of the century, though on a rather different basis of thought.

The use of alloy steels continued, both in such large structures as the East Bay cantilever, where over half the tonnage is of alloy or heat-treated steel, and in small highway bridges. The latter tendency is especially noteworthy. Investigations of the desirability of utilizing structural aluminum in special cases continue.

DYNAMIC STUDIES

Earthquake-resistant design received attention on the West Coast. To designers elsewhere this work is important in developing clearer understanding of continuous action in structures and especially in awakening interest in structural dynamics. It is literally true that the profession has been shaken out of its apathy in these fields.

In another way the work of J. B. Hunley for the American Railway Engineering Association is notable in the field of dynamics. His analytical approach stems from the work of Inglis in England, but the interpretation is distinctly American. His investigations and the work on earthquake resistance promise renewed attack on a problem that is so often raised only to be avoided: clearer definition of the significance of rigidity.

The American Railway Engineering Association adopted a new specification for steel railway bridges, previously presented for discussion. This is the first general revision in fifteen years. Notable clauses deal with increase in basic working stress, a new formula for stiffener spacing, revision of the rules for computing net section and new specified impact.

Timber again receives systematic attention. Tests of the resistance of timber floors to earthquake forces were made in California, tests of laminated floors of timber and of laminated timber arches at Illinois. Best of all, the West Coast built timber bridges with some of the grace of those of fifty years ago.

Brick, reinforced, once more promises to regain to some extent its position as a structural supporting material instead of a mere covering.

RISE OF THE RIGID-FRAME STRUCTURE

The structural type-of-the-year is definitely the rigid frame. The remarkable tests on concrete arches at Illinois are now followed by tests on rigid frames of concrete and on knees for rigid frames of various details of design. At the Bureau of Standards, tests on rigid frames of steel are in progress.

Definitely important are developments in design of highway bridges of moderate span. These were at one time, with industrial buildings, the most poorly designed and often the most poorly planned of our structural types. More and more the state highway bridge offices, cooperating with the Bureau of Public Roads, have broken away from standardized precedent and have become important organizations for research in design; new types appear, old types are improved, and special attention is given to important details of design. Continuous structures are more in use, deck girders supplant through girders, "in order to accentuate the roadway as a main-purpose element of profile." Excellent! This calls for a better type of engineer: pay him!

Notable is the increased use of rolled beams for spans of more than 100 ft. Important is the attention to the design of floors; very promising the rethinking of abutment and pier design; more important the attention to functional and esthetic design. The small highway bridge becomes again a problem in civil engineering and not a specialized problem in structural engineering.

In the analytical field, interest in analysis of continuous structures continues. Much of the work has shown acute obsession for fragments of knowledge, meticulous attention to details of computation, too little concern with unified concepts of structural action. On the whole, however, the activity is distinctly healthy, as the younger engineer develops a clearer picture of structural action; the work is valuable so far as it develops tools of thought rather than standardized types of thinking.

Studies by models and by polarized light continue to occupy our attention. At Lehigh, studies by polarized light were used to supplement laboratory tests of welded joints. Models continue to be used for dynamic studies at Stanford and elsewhere on the West Coast and at the Massachusetts Institute of Technology, and in general for problems varying from the erection procedure on the

Davenport Bridge to such quaint and curious academic tasks as that of drawing influence lines for the effect of transverse loads on a two-legged bent.

ORIGINALITY MARKS THE DAY

This depression period has been one of originality in architectural form and in structural type—rigid frames, shell domes, and in the state of Washington a relatively long bridge truss of unusual form in reinforced concrete. It has been even more a period of experiment of analysis, and of revision of concept.

The beginner wonders if the mass of theoretical and experimental material; the data as to new materials, new forms, new details can ever be assimilated into the structural organism. The seasoned practitioner rarely underestimates the digestive power of his art. During the past years, activity in design and construction has not been high; but rarely has the profession shown greater activity in reappraising its art, in revaluing its constructive thought. I think we still carry on sanely the tradition of the past four thousand years.

The optimistic tone of this article is striking. Cross was clearly aware that the Depression period was one of remarkable progress in the structural art. This is a point which often goes unrecognized, even today.

For American architects and structural engineers in the period from 1945 to 1960, it was fortuitous that the Cross method was at hand. During these years two forces were pushing the building industry toward the increased use of reinforced concrete for frames in multi-storied buildings. The first was the increased cost of steel. Though it was in every way a marvelous material for the skyscraper, its cost was rising during these years, and even affluent corporations began to consider the alternative of reinforced concrete. The second factor was an improvement in the technology of reinforced concrete. Pre-stressing and post-tensioning became common. Finally, the concrete industry itself improved the quality of its product.

A convenient example of Hardy Cross's impact is in the structural design of the Equitable Building by Pietro Belluschi (Portland, Oregon, 1948). The background for this handsome structure has been well discussed by Meredith Clausen. Suffice it to say that the ambition of both architect and client was to do a building which would be as technologically advanced as possible. For example, the architect made the maximum possible use of aluminum in the cladding. This material was also investigated for the structural frame but proved impractical. A steel frame was also considered, but steel was rejected both because of cost and the threat of adverse reaction between the aluminum sheathing and the steel. Hence the architect and client settled on reinforced concrete. With recent improvements this material now had strength of 7000 psi. Belluschi was therefore able to reduce the dimensions of the frame to an absolute minimum, even at the base. "This reduction," says Professor Clausen, "allowed a gain in space, both in the upper stories of offices and in the ground floor lobby and retailing quarters."[27]

Construction photographs make it absolutely clear that the building has a continuous rigid frame of reinforced concrete (Figures 12 and 13). The frame is, moreover, attenuated to such an extent that most casual observers think it is

Figure 12. Equitable Building, Portland, Oregon. Pietro Belluschi, 1947–48. Oregon Historical Society.

Figure 13. Equitable Building under construction, Portland, Oregon. Oregon Historical Society.

steel. The engineers on this building served Belluschi well. Perhaps it is worth noting that he himself was trained as an engineer and was in a position to push them to the utmost. In 1947–48, when the Equitable was erected, such a building would necessarily have been calculated with Hardy Cross's method.

What were the essential characteristics of Cross's mind? Nathan Newmark wrote that "Cross was not a particularly strong mathematician but he had a deep physical insight that led him to the appropriate mathematical solution of the problems he was concerned with in structural analysis."[28] I am forced to differ with this remark somewhat, since geometry is usually conceived to be a branch of mathematics. From all the internal evidence of his papers, Cross himself thought of geometry as the key to his work. We have alluded to his insistence on its employment in several of his papers. It was the key to his employment of graphic statics.

Cross was known as an engineer-philosopher, and his engineering thought certainly had an unusual intellectual scope. In his paper on "Limitations and Application of Structural Analysis" he quoted Sir Arthur Eddington approvingly: "No experiment is worthy of credence unless supported by an adequate theory." Cross agreed and added that not only is the adequate theory necessary after laboratory experimentation for illuminating the latter, but it is needed before the experiments are begun in order to plan the tests properly. The role of the testing laboratory for Cross has never been documented, but it is clear that it was extremely important. In his publications, his distinction was his discovery of the significance of convergence and of the iterative process.

In the United States we have produced several brilliant engineers who have been responsible for great structures like the Golden Gate Bridge, and a few like Gustav Eiffel, who have shown talent in both engineering and business. But in America Hardy Cross stands almost alone as an engineering theorist. Perhaps his only rival in the United States is Fazlur Khan. Although he is internationally known as a practitioner for the structural design of skyscrapers, particularly the Sears Tower and the John Hancock Center in which he used a tubular design of his own invention, Khan was also a theoretician. The amount of tedious calculation for statically indeterminate frames was reduced dramatically after Hardy Cross discovered the moment distribution method. However, exact two- and three-dimensional analyses became possible with the advent of the digital computer. Prior to that time, Cross's method was popular as an approximate procedure that was intended for beams and planar frames. All rigid-frame structures were analyzed as plane frames along the principal axes of the building. During the era of primitive computer analysis methods, Khan developed an approximate analysis method of framed, tubed buildings in three dimensions. He defined the terms *stiffness ratio, stiffness factor, shear lag,* and *aspect ratio* (somewhat similar to distribution factor, carry-over factor, etc.) for Cross's method as the parameters for such analysis. Khan further developed influence curves that could be used as design aids. Although these are not commonly used as design aids now, they are still appealing for manual calculation for preliminary design and for the design engineer who wants to obtain a good grasp of three-dimensional behavior of buildings in simple and intuitive terms. The way that Cross disposed of the conventional wisdom on continuity reminds us of the way that Khan gave up the conventional wisdom of using "rigid frame only" for tall buildings and developed a "systems chart" for high rise structures. Thus both men revolutionized the analysis of structures and the way that we view them. Both of them had an almost mystical understanding of structural behavior and pushed forward the state of the art. There is a remarkable parallelism here. Khan was the true heir to the engineering philosophy of Hardy Cross.

Both Cross and Khan realized that, unless one understands the behavior of a structure under various loading and/or imposed deformation situations, there is quite likely to be erroneous analysis which may result in faulty design. That these premises are correct in modern times can be verified by observing the serious and costly damage, in tens of billions of dollars, resulting from the 1994 Northridge and the 1995 Kobe earthquakes.

The analyst must take into account building behavior even in these extreme situations.

As we have noted, Cross was seen in his own time as an engineer-philosopher. We have pondered long over this somewhat peculiar designation. What does it really signify? In the classic sense a philosopher is a lover of knowledge who asks basic questions. One thinks of the magnificent Platonic injunction to Socrates: "Tell us, O Socrates, what it is that makes one thing a good and another thing an evil, whether seen or unseen, whether done or undone, by Gods or by Men." Hardy Cross was at home with this kind of thinking. His work shows that he thought that basic questions were the only kind worth asking. When he began, continuity was considered a problem. When he finished, it was viewed as an opportunity.

In a sense, Cross's approach was very American, if not Emersonian. His bibliography is full of papers which are defenses of hunches and intuitions. Of course, he assumed that the proper questions had been asked and that the computations were correct. But deep within his soul he responded to the famous Emersonian injunction "Trust thyself: Every heart vibrates to this iron string." He was also very much in line with the pragmatism of William James, who believed that truth was what worked.

And there is another sense in which Cross was an engineer-philosopher. In his approach to structure he curiously foreshadowed certain aspects of quantum mechanics. Norbert Wiener put it as follows:

> No scientific measurement can be expected to be completely accurate, nor can the results of any computation with inaccurate data be taken as precise. The traditional Newtonian physics takes inaccurate observations, gives them an accuracy which does not exist, computes the results to which they should lead, and then eases off the precision of these results on the basis of the inaccuracy of the original data. The modern attitude in physics departs from that of Newton in that it works with inaccurate data at the exact level of precision with which they will be observed and tries to compute the imperfectly accurate results without going through any stage at which the data are assumed to be perfectly known.[29]

Cross may not have known Einstein and Heisenberg, but his best work intuitively anticipates much of the philosophy of modern physics. In the conclusion of his paper on moment distribution, he seems to be grappling with Heisenberg's Uncertainty Principle. Since the values of moments and shears could not be found exactly, he would accept inaccurate data and admit that he was computing inaccurate results, which, as with Heisenberg, would be good enough to work with. A man who thinks at this level does indeed deserve the title of "engineer-philosopher."

3. The Years at Yale, 1937–51, and Retirement

> The East looked to Europe in matters of intellectual fashion, and in Europe the ancient aristocratic bias against manual labor lived on. Engineering was looked upon as nothing more than normal labor raised to the level of a science. There was pure science and there was engineering which was merely practical. Back East engineers ranked, socially, below lawyers, doctors, Army colonels, Navy captains; English, history, biology, chemistry, and physics professors; and business executives.
> —Tom Wolfe, Ph.D., Yale, 1957, *Hooking Up*

When Hardy Cross began his duties as chair of the Department of Civil Engineering at Yale University on July 1, 1937, he returned to a part of the country he knew well. He had, after all, taken degrees at M.I.T. and Harvard, and taught for several years at Brown. Part of the attraction for Cross, who had a considerable sense of history, may have been that he was filling a professional chair which was established in 1852 and was one of the oldest of its kind in the United States. He was only the fourth man to hold the post. The first was William Augustus Norton. An additional attraction was undoubtedly financial. The 1930s were depression years with generally tight budgets for Midwestern state universities.

From the point of view of Yale, the university was recruiting a man of substantial distinction. He had a tremendous reputation in the profession for his work at Illinois, especially for his method of solving statically indeterminate structures. He had two honorary degrees, one from Hampden-Sydney in 1934 and another from Lehigh University in 1937. He had received the Norman Medal of the American Society of Civil Engineers in 1933 and the Wason Medal of the American Concrete Institute in 1937. Still on the horizon was the Lamme Medal of the American Society of Engineering Education in 1944. Cross had also served on the National Committee on Suspension Bridges of the American Society of Engineering Education and on the technical committees of the American Society of Civil Engineers and the American Railway Engineering Association. Yale had attracted a man who was at the top of his profession in every way.

While at Yale, Cross participated in two important consultations. The first concerned the settlement of the Charity Hospital in New Orleans. Begun in July, 1937, the hospital showed alarming cracks in its façade in the winter of 1938 (Figure 14). The Board of Administrators turned to Dr. Karl Terzaghi, the internationally known expert on soil mechanics, and to Hardy Cross for advice on other questions. According to Terzaghi's biographer, he visited the city in March 1939, and quickly saw that the difficulty was a consolidation settlement. It was mostly completed, and it agreed with his theories. With regard to the building's safety, he thought that it was probably no worse than most buildings on clay, many of which suffer higher maximum stresses in the steel skeleton than the designer anticipates. However, since the building was unusual and unprecedented in several respects, he recommended re-analysis by Hardy Cross, presumably to see if the settlement had affected the frame.[1] Whether or not Cross owed his appointment to Terzaghi is an open question. In any event, it was Cross whose report was published in *The Times Picayune* on Wednesday, July 19, 1939. The author mentions Terzaghi nowhere in the report. This omission is characteristic. Cross, a temperamental genius, did not really work well with the best of his contemporaries. He wrote:

> The results of the investigation justify the following conclusions:
> 1. It can be stated with the limitations indicated in conclusions number 5 and number 6 below, that immediate occupancy of the building is justified.
> 2. The obvious cracking in the walls and partitions has no relation to the safety of the structure. These cracks show only that settlement has occurred, a fact already known in other ways. Determination of the effect of this settlement on the safety of the structural framework is a matter for the separate study here presented.
> 3. Most connections used in this structure and the properties of the steel permit large deformations without serious danger.
> 4. No remedial changes in the structural framework are recommended except in the columns supporting the girders above the chapel.
> 5. While these columns are probably not in danger, some steps should, as a matter of engineering precaution, be taken to improve their condition.
> 6. This report refers to present conditions. It does not discuss causes or progress of settlement nor anticipate changes in conditions of the superstructure resulting from further settlement. Hence it is recommended that observations at critical points be continued for several years.

INVESTIGATION DATA

The writer was first requested to report on the strength of the superstructure of this building on March 9, 1939. Since then he has personally examined the structure, both the obvious cracking and also, at various points, the condition of the floors and of the enclosed steel work which supports floors and walls of the building. He has at various times been supplied by the architects of the building and their structural engineer with the original plans and design computations, and with supplementary data dealing with displacement. These data have every evidence of dependability.

The writer has examined the plans and, so far as necessary for the purpose of this report, the design computations. This examination shows that the building was designed in conformance to the best practice and that the plans were carefully prepared to give a satisfactory structure based on expectation of ordinary settlements. What changes would have been desirable in the plans of the superstructure had the differential settlement that has occurred been foreseen, is not here discussed though it is an interesting problem to which the answer is not obvious.

Reports of mill inspection show that the steel conforms to standard requirements. The steel that could reasonably be inspected in the completed structure shows good workmanship.

Unusual cracking of interior partitions, of plaster work, and of the stone work in the outer wall is at once apparent, in the lower three stories of the building. There is some interior cracking in nearly all stories. These partitions and walls serve no structural function. The cracking is unsightly, but does not impair the strength of the building. There is also some cracking, not unusual in buildings, in the basement walls.

CONDITION OF THE STRUCTURAL FRAME

On the basis of formulas and criteria commonly used in the design of buildings, there can be no doubt that portions of this building are to various degrees overstrained as a result of settlement. I have examined the problem of overstrain with special reference to the following questions:

First. Possibility of overloading column from vertical shearing induced by walls and partitions.

Overloading of the columns because the shears from partitions or walls is considered unimportant because of:

(a) Low strength of the masonry in the outer walls to produce such shears, especially at the time when most of the differential settlement occurred.

(b) Negligible strength of this type of action of the interior partitions.

(c) The type of architectural details used in the walls above the third story.

(d) The fact that any such possibility is largely eliminated after the walls or partitions are cracked.

Second. Stresses in the floors in their own plane resulting from unequal horizontal displacements of portions of any given floor.

Investigation of this question shows that the effect is not important.

Third. Condition of the joints of the structural frame.

I have examined typical joints in the steel structure, some of which are especially likely to show large deformation. None of these shows distress. The rivets are tight and the connections and clip angles do not show deformation. This shows that the joints are at present fully able to carry their load.

Fourth. Strength of the beams and of the floor joints.

The strength of these elements of the framework is not impaired by the settlement.

Fifth. Flexural strain in columnar in combination with axial stress.

The columns of this building are subject to a combination of flexural strain with axial stress. It is to be recognized that the tendency of girders to produce bending in the columns to which they are connected exists in all buildings. Under the loads acting on the girders, the ends of the girders rotate and especially at the walls, rotate the columns to which they are attached. Moreover, these rotations due to load are often of a relatively high order of magnitude, fully comparable to those resulting from settlement in many parts of this building.

The flexure of columns by connecting girders may be relieved and commonly is relieved by several phenomena. We note:

(a) Deformation of the connection angles, of the rivets and other participating parts at junctions of columns and girders.
(b) Sometimes where connections are eccentric to the center line of the column, a light twisting of the column avoids appreciable bending.
(c) For large moments where the relative proportions of the members is of the order common in this building, shearing distortion of the column web within the joint without appreciable weakening of the column prevents high flexure in the column.
(d) In some joints, the connection details are such that dangerous moments could not be transmitted to the columns.

The nature of the connections commonly used in steel work and the plasticity of steel make possible large deformations without damage. This plasticity may, under certain conditions, result in buckling. For this reason, as well as because of their importance as structural members, special attention has been given to studies of the columns.

During the investigation computations have been made, the literature has been examined, reports of laboratory tests have been scrutinized and experiences and observations in other cases have been studied. However, as is usual in practical matters, the conclusion in each particular case results from a judicious synthesis of the evidence from various sources. It is unnecessary to itemize in each case the reasons that have led to my conclusions.

My conclusion is that the structural framework is in a safe condition, with the limiting considerations applicable to the four columns discussed below.

SPECIAL THEORY OF THE ACTION OF COLUMNS DEFORMED AS IN THIS STRUCTURE

I have developed a theory as to the probability of buckling in columns such as these from end rotations of the nature here under consideration. No matter how great these rotations may be, provided the axial stresses are of the order here considered, this theory indicates great improbability of buckling in columns such as we have in this building. This theory is, so far as I know, without direct support from published data nor is it supported by clearly applicable experience from the field. Since the theory is a result of abstract reasoning without independent support, cautious engineering does not justify depending on it alone in so important a structure.

COLUMNS CARRYING THE GIRDERS ABOVE THE CHAPEL

Notwithstanding my theory of buckling applicable to columns in this building, I am unwilling to assume the responsibility for approving the columns supporting the main girders above the chapel unless steps are taken to ameliorate the condition of overstress in them. Because of the theory of column action referred to above, I do not expect collapse of these columns—this should be clearly understood. Since, however (1) this theory appears to be without support except from my own reasoning, (2) the stability of these columns is so important, and (3) there is no good reason to suppose that flexural strains in these columns are relieved in other ways, I advise that you seriously consider improving the condition of overstress in these columns between the second and third floors. This can be done in several ways. (These columns are referred to on the plans as columns 49E, 49W, 51E, 51W).

FUTURE OBSERVATIONS RECOMMENDED

Because this report is based on the condition of the structure at the present time I strongly recommend that observations at critical points be made at intervals somewhat frequent at present, and continuing with less frequency over a period of several years.

For students of Cross's work this report is invaluable. It is the only document we have which shows him at work in the gritty world of builders and politically connected architects. The designers of the hospital were the local

Figure 14. Charity Hospital, New Orleans, Louisiana. Williams Library, the Historic New Orleans Collection.

firm of Weiss, Dreyfous, and Seiferth, which had strong ties to the Huey Long machine, still operating despite the assassination of its leader in 1935. The hospital itself was a $12 million project which had distinct resemblances to the New York Hospital, completed a few years earlier. The general contractor was the large and well-reputed firm of George A. Fuller & Co. The hospital had been held up for several years because of the enmity between Long and Franklin D. Roosevelt. After Long's death, W.P.A. funds became available. The reason for the differential settlement must have been known to a great many people in New Orleans. Of 9567 pilings used to support the hospital, 8019 were to have extended 43 feet into the ground. In point of fact, only 7,532 long pilings were used, and they measured only 28 feet. Dr. Abe Mickal, former L.S.U. football star and a medical student at the time, watched the pilings disappear into the mud when the pile driver hit them. According to a recent historian, the length of the pilings was "substantially less than the length of pilings under any comparable building constructed in New Orleans in recent years."[2] Hence the Board of Administrators confronted the following settlement figures:

8/15/39	12.844"
9/25/39	12.96"
2/15/40	14.20"
2/15/41	15.00"
8/24/42	17.16"
4/8/43	17.86"

The solution delicately alluded to by Cross was actually quite obvious but not at all simple. It would have involved elimination of the corruption from major public building projects in Louisiana.

Cross, however, confined himself to the relationship between the settling and the structural frame of the hospital. With regard to this problem, he found reason to recommend only minor repairs and continuing inspections of four columns. He himself had a theory of buckling which strongly indicated the safety of the four columns, but since he did not have empirical proof, he thought that cautious engineering practice dictated reinforcement at these four points. Time has vindicated his judgment. In 1992 Dr. John Salvaggio wrote, "Despite the rumors, the hospital was, and remains to this date, a sound physical structure, and its future has in no way been jeopardized by the settling problem."[3]

The unhappy conclusion of the consultancy was that thereafter Cross was on bad terms with Terzaghi. Perhaps the quiet and refined southerner automatically disliked the somewhat flamboyant European. Terzaghi's biographer notes that Terzaghi reciprocated the feeling, as well as Cross's propensity for omitting the names of his collaborators. Terzaghi did not even reference the work of Dimitri Krynine, Professor of Soil Mechanics at Yale, in his own massive publications on the subject. For most historians, Karl Terzaghi (1883–1963) is the founder of the engineering science of soil mechanics. Born into a military family in the Italian portion of the Austro-Hungarian empire, he attended the Technische Hochschule in Graz and served in the First World War. From 1918 to 1922 he taught at Robert College in Istanbul and there conceived his basic work, *Erdbaumechanik auf bodenphysikalisches Grundlage.* This was soon translated into English and appeared as a series of essays in the *Engineering News Record.* Terzaghi had to fight many battles for his new discipline. This task he undertook with zest, since he was, " . . . a brilliant man of vast energy and huge self-confidence," according to a contemporary. In the United States, the acceptance of soil mechanics as an engineering science may probably be dated from his first appointment at Harvard University in 1939. Interestingly, Terzaghi had some of the same reservations about mathematics in engineering as Cross. He was afraid that an overly great reliance on mathematics might obscure physical reality.

The other consultation in which Cross played a leading role was on the collapse of "Galloping Gertie," the famous Tacoma Narrows Bridge, on November 7, 1940 (Figure 15). This structural failure was particularly well-documented. From its completion in early summer of that year, observers noted pronounced vertical oscillations in even the lightest wind. Automobile passengers became seasick. It was not unusual for suspension bridges to show some amount of movement. The Golden Gate Bridge in San Francisco had moved up and down noticeably in gale winds of sixty mph and up to six feet laterally in other high winds. The main difference between the Tacoma oscillations and those of other suspension bridges was that elsewhere the oscillations damped (died down) quickly. At Tacoma the movements seemed to last forever. Engineers were sufficiently worried that Professor F. B. Farquharson at the University of Washington decided to build a large-scale model and to monitor the bridge and movies while studying the problem on the model. His film on the failure of the bridge is familiar to every one who is interested in such structures.

Figure 15. Tacoma Narrows Bridge. November 7, 1940. Special Collections, University Archives Division, University of Washington Libraries, no. 290.

The Years at Yale and Retirement   61

So spectacular was the Tacoma Narrows collapse that President Roosevelt named a national committee to study it. Othmar Amman, designer of numerous successful suspension bridges in the East, was the chairman. Other members were Hardy Cross and the aeronautical engineer Theodore von Karman. Initially, von Karman was annoyed with the civil engineers. "Their thinking," he said, "was still largely influenced by consideration of 'static forces' like weight and pressure which create no motion instead of dynamic forces which produce motion or changes of motion. Bridges had been observed to oscillate in wind before, but nobody had thought such motion was important."[4]

Von Karman (1881–1964) was the best aeronautical engineer of his time. He is often called the "father of the supersonic age"; his analyses of the action of fluids led, among other things, to the development of efficient aircraft structures, of wind and shock tunnels for testing aircraft, and of supersonic jets, guided missiles, and rockets. He was director of the aeronautics center at California Institute of Technology from 1930 to 1949 and founder and chairman of the advisory council for air research and development of the North Atlantic Treaty Organization from 1951 to 1963.

Von Karman reports one sharp controversy during the investigation of the Tacoma Narrows Bridge. Othmar Amman, who was thinking in terms of static load, refused to recognize that small loads can be very dangerous when they cause repeated oscillation. Ultimately the theory of the von Karman vortices won out. Matthys Levy and Mario Salvadori accept it and in their work show a diagram of wind-induced twist.[5] More recently Henry Petroski notes the failure of Leon Moissieff, the designer, to take account of the previous failures of suspension bridges.[6] Cross's role on the committee is not recorded. We do know, from the notes of Thomas Keusel, that Cross was intensely interested in suspension bridges.

The effect of Cross's work in the United States can be seen clearly in a passage by David Steinman on "The Bridge Engineer as a Mathematician." In 1941, Steinman, who was himself one of the great bridge builders of the age, observed that the mathematical theory of bridge design had been vastly extended in recent years. Precise analysis had replaced rule of thumb. The new procedures meant stronger and more efficient structures with a smaller expenditure on material. If the recently developed mathematical analysis had been available twenty years ago, as much as sixty percent savings could have been made in certain structures along with an increase in their safety. Previously, statically indeterminate forms had been avoided because they were difficult to analyze. Now rigid frames, continuous spans, hingeless arches, and other indeterminate structures, had come into general use as their methods of analysis and computation had been improved, simplified, and perfected. Without any mention of Hardy Cross, Steinman's debt to Cross is perfectly clear. Perhaps because of the Tacoma Narrows disaster, he noted that a knowledge of aerodynamics would in the future be a requirement for any successful bridge engineer.[7]

Steinman himself built almost entirely in steel, but he could not avoid taking notice of the popularity of reinforced concrete in new forms, particularly pre-cast and pre-stressed beams. Pre-casting offered an obvious saving in formwork, elimination of much on-the-job labor, speed of construction, and closer, more accurate control of the concrete mix. Steinman believed that Eugene Freyssinet in France and Gustav Magnel in Belgium were the joint fathers of pre-stressing. Today Freyssinet is probably the better known of the two partly because he was picked up by various art historians, notably Siegfried Giedion, but in his own time Magnel was almost as important. From 1946 to 1947, Magnel was in the United States, partly for his health and partly to do some experimental work at the University of Pennsylvania. Cross scheduled him as a guest lecturer in his course (Magnel spoke fluent English—he had spent the World War I in London). The Walnut Lane Bridge in Philadelphia, completed in 1951, was a large, important structure of thirteen 160–foot girders and fourteen 74–foot ones. The method of pre-stressing was Magnel's. Freyssinet had less impact in the United States, though the gifted engineer Conde McCullough used his method of hydraulic jacks at the Isaac Lee Paterson Bridge at Gold Beach, Oregon, in 1932. Cross was very interested in the technology of pre-stressing and post-tensioning, techniques which began to appear near the end of his academic career.

Significant international recognition came to Cross in European journals shortly after the publication of his important works in the early 1930s. In England, a writer in the November 24, 1933, issue of *Engineering: A Weekly Journal*, remarked, "No technical paper of recent years has drawn forth such an extended discussion or received so many favorable comments as 'Analysis of Continuous Frames by Distributing Fixed End Moments' by Hardy Cross, Professor of Structural Engineering, University of Illinois." The writer observed that the paper was brief and concise—too much so for many structural engineers—and went on to discuss the commentaries on the Cross method which were already beginning to appear on both sides of the Atlantic. He was not altogether enthusiastic. The Cross method had its limitations, and the calculations for tall building frames could be tedious. Nonetheless the procedure was a vast improvement over anything previously known.

The authors of British textbooks on civil engineering in the 1930s and 1950s gave substantial space to the Cross procedure. Pippard and Baker devoted several pages to Cross in their *Analysis of Engineering Structures* (London, 1936). Several years after World War II, H. W. Coulter wrote:

> The method may be modified in many ways and offers a number of short cuts in specific problems. Once the general principles of the method are understood, the student will have acquired a point of view which should enable him to arrive at a reasonable solution of any statically indeterminate problem involving a continuous type of structure, where the moments are the controlling factor in the design without the necessity of any mathematical work except the simplest arithmetic.[8]

In Great Britain, as in the United States, the Cross method remained important until the advent of the digital computer.

The first notice of Cross's work in French was a series of articles by Serge Zaytzeff in *La Technique Moderne—Construction* in 1947, 1948, and 1950. These articles were gathered together and published by Dunod in Paris in 1951 under the title *La Methode de Hardy Cross et Ses Simplifications.* In the preface Zaytzeff noted the difficulties of calculating indeterminate structures and declared that the appearance of Cross's method was an event of capital importance. This short, 80–page book was so successful that the author expanded it to 224 pages in a 1953 second edition. Both works were not so much translations of Cross as a series of commentaries with suggestions for short cuts. This result was, I think, exactly what Cross wanted. He had deliberately written his initial paper in terms which invited short cuts and emendations. His responses to the original publication make this point very clearly. He was probably delighted to see Zaytzeff's application of the Cross theorems to the Vierendeel truss in his second edition. His original paper has, as I have indicated, a long list of subjects to which his method might be applied. The Vierendeel truss was ripe for investigation.

It was in the German-speaking world that the reception of Cross's work was really enthusiastic. The first notice was an article by M. F. Fornerod of Zurich on pages 223–27 of *Der Schweizerische Bauzeitung* of November 4, 1933. In a sense the early notice of Cross's achievement in this periodical is not surprising. *Der Schweizerische Bauzeitung* is a journal which has always paid a good deal of attention to American developments. In 1912 the editors published important articles by H. P. Berlage on Frank Lloyd Wright. In 1926 they carried the criticism of Robert Maillart on the flat slab system of Claude A. P. Turner, so it had been attentive to American issues for a long time. Fornerod had an excellent grasp on the Cross method. He saw the importance of distributing the fixed-end moment at every joint and of the convergence obtained by successive approximations. He also noted the special utility of the procedure for multi-story buildings subject to wind load and gave as an example the calculations used for the twenty-two-story American Insurance Union in New York City. His verdict was that a great deal of time was saved by the introduction of the Cross method. Fornerod entitled his article, "Calculation of Continuous Frames by the Method of Moment Distribution," and it marked the beginning of the appreciation of Cross in the German-speaking lands.

Five years later, on January 21, 1938, in *Der Bauingenieur,* pp. 45–52, came "Improved Methods of Calculation for Continuous Beams and Frames" by W. Dernedde of Dortmund. A provocative article, it is interesting because it deals with the emendations and short cuts achieved by the engineering community in only a few short years. Dernedde pays particular attention to the improvement made by T. Y. Lin in 1934. Cross himself must have been delighted with this piece.

After the Second World War there were at least twenty additional commentaries in German on Cross's work, mostly from Berlin, but a few from Vienna and Budapest. The majority were published by Springer-Verlag, which had branches everywhere. Typical of these publications was the formidably titled, *Des Verwollstandigte Cross—Verfahren in der Rahmen Berechnung.* (The Complete Cross Method in the Calculation of Frames.) This volume was published in 1962 in Berlin, Göttingen, and Heidelberg. It was a revision and expansion of an earlier work, *Das Cross-Verfahren* by Johannes Johannson, first published in 1948 with a second edition in 1954, with the assistance of Engineer Gunter Raczat of Hagen in Westphalia. The earlier editions had been essentially handbooks for practicing engineers. Raczat proposed to expand its theoretical base and take into account the possibility of electronic calculations. He offered the best German assessment of Cross's contribution this writer has encountered. The American engineer had, said Raczat, caused a revolution in the area of the statics of frames. He added that the "echo" of Cross's paper of 1930 was comparable to only a few technical studies. It had inspired a tremendous amount of fruitful work by other engineers. The paper had, in fact, brought about a whole new view of engineering. About 1930 there were innumerable rules for the solution of statical indeterminacy, Raczat noted. These rules led to much confusion. The Cross method offered a way out of this troublesome situation. Cross used only a few unknowns in his solution, and the work in which he described his idea was very short and wholly limited to his basic concept. The possible number of changes and variants was only suggested. Raczat called this tactic a masterstroke of pedagogy and thought that it was responsible for the worldwide adoption of his ideas. The German author noted that Cross's work was greeted enthusiastically by those who were not mathematically minded and that it was a help to the mathematicians as well. He spoke of a wave of insights which followed the initial publication and specifically mentioned the American engineer Lawrence Grinter. The method made the necessary iterations simple and easy to complete. In the old textbooks there were piles of formulas and ways of measuring stiffness. Everywhere there were rules for the acceptance of zero moments and methods which demanded a high level of mathematical skill. Cross swept all this away. Raczat saw his method as a manifestation of the empirical inclination of the Anglo-Saxon peoples and their pragmatic way of thinking. He quoted approvingly Cross's remark that since the values of moments and shears could not be found exactly, it was senseless to attempt to do so. Finally he noted that in 1951 Cross had begun his retirement and that he had had the pleasure of seeing his work recognized in the entire world.

In addition to the items in French and German, the bibliography of writings on Cross lists works in Italian, Spanish, Dutch, and Portuguese. But I must add, in the manner of Cross himself, that, if I could obtain these works, I could not evaluate them because I do not read Spanish, Italian, Dutch, or Portuguese. It is clear, however, that Cross's work showed a new maturity in American structural engineering. Until his time the theoretical basis of the profession had been largely developed by Europeans. In Hardy Cross the United States produced an engineer-scientist who built on European developments and worked out solutions which achieved worldwide recognition. This is a mark of true cultural maturity in American thought in engineering.

Cross and his wife traveled a good deal in Europe, but we do not know if he called on any of the men who wrote books about him. Considering his personal reticence and his deafness, it seems unlikely. Cross was, however, an extremely knowledgeable tourist, and was particularly fond of medieval architecture. In a memorable passage in his essay on standardization he observed:

> Medieval architecture was not standardized. That is one of its great charms. Dissymmetry is marked; apparently it is frequently intentional in the medieval cathedral. There is nothing very standard about Chartres and Mont St. Michel. The little naked soul so prominent in sculptures of the Day of Judgment did not always outweigh the devil and his imps; in one of the column capitals of Saint Lo, the sculptor, perhaps suffering from morbid indigestion, reversed the procedure and thus caused great embarrassment for future curators.[9]

It appears that Cross was a close observer of every detail in a medieval building. His comments on medieval architecture are part of a trenchant passage on standardization. In civil engineering, he notes, standardization had been carried far and that for reinforced concrete, elaborate standards had been set up. In fact, there was even a "standardized theory of reinforced concrete." In Cross's view, however, this theory was "as complicated a bit of nonsense as has been conceived by the human mind."[10]

It was mainly useful as a check on the non-discriminating intelligence. Cross could, on occasion, be quite acerbic—and he took a very long view of the history of engineering. He held that it was almost impossible to put dates on engineering, and that it is equally hard to say that there are entirely new problems. The problems of the present were in many respects the problems of hundreds of years ago, but contemporary issues often involved new materials and always, different conditions.

In many respects Cross comes across as conservative in his attitudes. Notwithstanding his own contributions, which were certainly revolutionary, he was constantly on the lookout for specious novelty in all fields: art, philosophy, economics, and religion. The claim of novelty, he thought, was used to cover up error and enliven dullness. He added that often a clear and direct restatement of a fundamental principle could have a profound effect. Taking a long view, Cross remarked that novelty and uniqueness were dependent on the conditions of the times. "Long timber trusses," he wrote, "are more news today than they were in 1850. Brunel used reinforced brickwork over a hundred years ago; the use of mechanical models is not by any means new."[11] And he thought that soil mechanics was simply a fancy name for the study of foundations, which had been going on for a long time. For Cross, the present—the 1940s—was a time to "take stock of what we know, what we do not know and what we need to know and why."[12] It would be interesting to know what Hardy Cross would make of the tremendous advances in engineering and science of the last half century.

Hardy Cross was a man accustomed to making sharp distinctions, and the third chapter of *Engineers and Ivory Towers* is devoted to discrimination between education, training, and schooling. He thought that the purpose of education was to prepare the whole person to live a full life in a whole world. This was a large order and seldom fully attained. Clearly Cross was against the narrowness so common in education for the engineering profession. It is clear that Cross believed that he himself had benefited by the remarkably thorough liberal education which he had enjoyed. Perhaps he would have liked to require something similar for everyone in a professional school, whether it was law, medicine, or engineering. Schooling, he thought, could help in guiding the student to information, but there were problems in the rigid disciplinary distinctions between departments of a university. Cross fulminated against professors who wanted to teach rules rather than judgment, though he admitted the necessity of teaching rules.

Cross had strong ideas on the nature of a university. He placed emphasis on the faculty and its work, on the campus life of student societies, and finally on the sense of community which was a great part of academic tradition. Hardy Cross was very much a part of the academic community at Brown, at Illinois, and at Yale. Finally, he was well aware that, "Too many old men, set in their ways, are ready to guide. And to guide often means to rule, to suppress, to kill."[13] Cross was particularly hard on those who would turn engineering colleges into trade schools. He was intensely critical of his own profession of engineering education and insisted that the function of universities was to turn out intelligent men with some knowledge of practical fields rather than dull practitioners with detailed knowledge of limited fields. He deplored the growing obsession with academic credits and the confusion between literacy, training, learning, and wisdom. He wanted students who would think for themselves and arrive at their own conclusions (like Alford). Cross could be devastating in his critiques: "It may be accepted that some bad education is worse than none and more bad education is worse than less. This needs to be stated and restated."[14]

Himself, a liberally educated man if ever there was one, Hardy Cross wanted his engineering students to be liberally educated men too. (It is unlikely that he ever imagined women in the engineering sciences. As this is written, Smith College has only recently established a school of engineering). He was, however, not much interested in requiring more courses in sociology, economics, psychology, or literature. All these disciplines, said Cross, appear when engineering principles are applied to the study of engineering works. If the undergraduate teaching of engineering was good enough, students would be stimulated to find out for themselves. Cross was very clear that teaching was an art, not a science. He excoriated the idea that all human activities can be mastered by the methods of the physical sciences. What Hardy Cross would have thought of the present obsession of education with computer science can only be guessed. He was fond of an anecdote about Raphael. A novice is supposed to have asked the artist with what he mixed his paints. The master replied "with brains." Whether his paint had a high or low Brinnell number was of no account.

"Teachers," said Cross, "have two responsibilities to their students: one to give them enough information and vocational education to enable them to get a job and to hold it till they get rooted in a highly competitive world, and the other, to train them in methods of thinking and investigation to meet the demands of an ever-changing world—demands the details for which none can foresee."[15] Cross himself did his best to teach judgment because he thought that it was a quality which would stay in demand. A big part of the education of an engineer was the development of a sensitivity to questions of proportion, scale, and setting.

In the December 1931 issue of *The Technograph*, the periodical published by the College of Engineering at the University of Illinois, Cross offered the following essay entitled, "Bridges Here and There":

> A room popular for years in Chicago's Sherman House bore in Gaelic a mural description, "Here's to the bridge that carries us over." That's what a bridge is for, to carry the roadway over some obstruction; but it may do it in a great many ways. The bridge is a part of the roadway, and it is also a part of the landscape, and a part of the river or valley which it crosses. It must harmonize with its environment; it must meet the spirit of its associates. In a park it may be a jolly little bridge, and play, as a little suspension bridge over the lake seems to play in the public gardens of Boston, but it must be very serious-minded where it is to carry a railway over a gorge. If it lives in pine forests, the bridge will perhaps want to be of timber and feel that it fits into the neighborhood, but we connect rock gorges with cut-stone masonry or with concrete, and for huge spans we use the strength and grace of steel.
>
> Paris, with all its fascination, center of art, ancient seat of learning, city of great vistas, of magnificent gardens, is also a city of beautiful bridges. Artist, architect, and engineer find fascination along and between the banks of the Seine. Pont Alexandre, Pont de la Concorde, Pont Royal, and all the bridges connecting the island with the banks fit gracefully and harmoniously into those magnificent vistas which stretch from Notre Dame to the Trocadero and from the Chamber of Deputies to the Madeleine. Pont de la Concorde, chef d'oeuvre of Perronet, architect to the king of France and first chief of engineers of the Department of Roads and Bridges, was being widened last year and I was able to examine the excellence of the workmanship throughout this fine structure. Perronet's notes are said to have contained plans worked out in detail for a masonry arch of a span length of 500 feet, a span which we often think of today as near the limit for reinforced concrete in spite of our fancied progress with theories of elasticity. This is about the span of the great arches recently constructed at Plougastel after designs by Freyssinet, and is more than one-half that of our Hell Gate bridge.
>
> At Paris, as elsewhere in Europe, the accumulation of beautiful bridges has been through a process of long selection. The beautiful bridge is a bridge well designed; a bridge well designed is, in general, durable. As the years go by, it becomes part of the life and affections of the people, a part of a city, focus for civic development. It captivates the fancy of artists and poets, and so endears itself that it is permitted to survive with small change as the years pass. American books often convey the impression that Europe has more numerous ex-

amples of the quaint and the beautiful in bridge architecture than has our own country. In so far as Europe has been able to preserve the best of its ancient bridges, that is true, but their more recent bridge architecture is, I believe, not superior to our own. You may see this in the newer bridges over the Seine and the Marne in the zone devastated by war. At Chateau Thierry, for example, the modern bridge of reinforced concrete seems mediocre and harmonizes little with the ancient buildings along the river or with the mouldering castle on the heights; we feel a little sympathy for this new material forced into such ancient and distinguished company.

Europe has few bridges which we would call large; the bridge over the Elbe at Hamburg and the Forth bridge are among the few which would by their size alone attract attention in our technical press. To these we may add a few over the Rhine and perhaps over the Danube. The bridge at Hamburg, consisting of two sinusoids intersecting at the piers and entered through massive and rather purposeless gate towers at the banks, fascinates by its curiousness rather than by its beauty. Forth squats spraddle-legged in the Firth like an antediluvian dinosaur, magnificent in size, but not distinguished in proportions. America is the home of the great bridge.

Bridges present one face to river travelers, another to travelers approaching by land, and a third to those who loiter by the parapets to fish or rest or dream. Much fine art has gone into the study of approaches, of pier forms, of details of balustrade. Each bridge has its own environment; it may be merely an extension of the street and be dominated by neighboring buildings, as is the case of Ponte S. Trinita; or it may itself dominate the view as does Risorgimento.

I have included three foreign views, not because America does not furnish beautiful examples, but because those given are less familiar to American readers.

* * *

Every fine bridge has a personality of its own. The Lars Anderson bridge in Cambridge, Massachusetts, charms with its companionship with the river; the bridges of Venice are part of that glorious ensemble of Renaissance architecture; the Eads bridge has grace and strength of line in keeping with the dignity of Father of Waters; the Charles bridge, over the Moldau at Prague, fascinates with its Jew's Cross and other fine statuary; some play in the park, some majestically span great rivers, but the bridges that impress themselves on the imagination fit their environment. Ponte Vecchio is charming over the Arno, Ponte de Rialto is part of the Grand Canal, but the Chicago river is another stream with another tempo.

Europe seems to have loved its rivers and its bridges more than we have learned to love ours. The embankments of the Seine, of the Thames, of the Tiber, the Alster Basin at Hamburg,—I think of quite as many lovely bridges at home as abroad, but abroad the river banks have been more fully developed to charm and rest those who pause to enjoy the hospitality of the bridges. We have appreciated the rivers in our own cities much less than we should have done and less than we will do in the future. Boston has done marvelously with her Charles River basin, Chicago is changing its river from a great sewer to a ribbon of restfulness, Pittsburgh is discovering the waterways whose junction made her a trading post, and Indianapolis has found that the White may be a thing of beauty as well as a flood maker.

* * *

This little essay may serve for some readers as introduction to a fascinating sideline of engineering. Nearly all of us are given to some hobby of collecting. Most undergraduates have a passion for collecting formulas and I find graduate students much given to collecting all sorts of variations of methods of analysis. Both varieties of birds-egging are likely to become vicious and for my civil engineer friends I recommend, as an outlet for postage stamp proclivities, an excursion in the field of bridge collecting. By photographs and descriptions, accumulation of historic associations and of artistic detail, one can build up a museum which is not only of interest as a hobby but also has value as a background for professional work. It is a real pleasure to turn from the exact mathematics of analysis or the details of connections to a more general view of the function of bridge structures. Look over the pictures in "A Book of Bridges" and "The Bridge," illustrated by the distinguished artist Frank Brangwyn, in our library; here you will see bridges, not as formulas, but as studies in light and shadow. Look over the fine collections of photography by Charles R. Whitney and Wilbur J. Wilson, and do not neglect the interesting collection of photographs of American bridges which Professor Shedd has put on the walls of engineering hall.

America today is developing excellent standards in bridge architecture. In the past we have been so busy building bridges that we have sometimes forgotten that they should be beautiful as well as useful. But where a bridge has been "right," that is, of materials which obviously fit into the community and of design which is structurally correct, our bridges have a dignity not to be surpassed in Europe. I think I have never received a greater thrill from the beauty of any bridge than from Walnut Lane arch when I first saw it from the bed of Wissahickon Creek.

A bridge must be structurally sound, correct in form, adequate in detail, of good materials, properly used, but it should also fit into the picture and perform in a graceful and dignified way its function of carrying the roadway over from street to street or from hill to hill. Remember that the distinction between architect and engineer is quite recent and that in bridge architecture it is almost impossible to enforce it. One who would design a beautiful bridge must have correct conceptions of structural action; the artist must be something of an engineer, the engineer something of an artist and city planner.

* * *

In this brief but delightful essay, we note that Cross shows a familiarity with the work of Eugene Freyssinet, the great French pioneer of reinforced concrete in the twentieth century.[16] Freyssinet (1879–1963) was highly interested in long span construction and in pre-stressed concrete. On the other hand, Cross holds no particular brief for modern work. He seems to prefer those bridges which have proved themselves over time. He was also extremely conscious of the importance of the urban setting within which the bridge engineer had to work. In later years he came to believe that America would become the land of great bridges. He would probably have been surprised by the great cable stay bridges over the Rhine and the Tagus and wondered why the United States could not emulate them.

* * *

Thus Yale had brought in a man who had an international reputation. Their new professor had strong views on the nature of engineering, on engineering education, and on the place of engineering in society. By virtue of the success which he had achieved, he did a certain amount of public speaking, and evidently he used his opportunities to set forth his views on these questions. There were acceptance speeches for his medals and public appearances on other occasions. Always a conscientious man, Cross inevitably wrote out his remarks. It was these writings which Robert C. Goodpasture studied and which were the basis of *Engineers and Ivory Towers*. We have been examining selections from the edited public utterances of a man who for several years was a major spokesperson for the profession of civil engineering. We should all be grateful to Mr. Goodpasture for recovering them. They are, however, necessarily somewhat pieced together. In his introduction, Goodpasture tells us that he used probably one-sixth of what was available. We have quoted extensively from his selection. The reason is a desire to give some sense of the flavor of the man's writing. *Engineers and Ivory Towers* is a work that is as pertinent to problems in engineering education today as it was when it appeared.

What sort of department chair was Hardy Cross? Immediately we can say that he was brilliant and demanding of both his faculty and his students. He had high standards for himself, and he attempted to impress these standards on everyone around him. At the same time, the available witnesses affirm that he had always the courteous bearing of a Southern gentleman. For his faculty he placed primary emphasis on teaching, and was well aware that good teaching took many forms. A man who was an excellent lecturer might be a washout in seminar. The converse could, of course, also be true. Cross expected that research and scholarship would be by-products of good teaching. He wanted teachers with lively minds who would ask lots of questions about their subjects. Cross did not mind if sometimes these were foolish questions. A university, he said, was a place to make many intellectual mistakes and learn how to rectify them. This is an unusual but useful definition of a university.

Of his students he expected as much as of his faculty and graduates, and he was a hard taskmaster who demanded rigorous attention during lectures and in classroom discussions. There was no possibility of drifting off to sleep in the classes of HX (thus he signed himself) no matter how late one had studied the night before. A man who studied with him at Yale recalls:

> Professor Cross was a clear and logical speaker as well as thinker. He chose simple words and simple illustrations to make his points. He emphasized the importance of visualizing structures. This included always making elementary sketches of structures under consideration. Then, he would require us to show the logical deflections of the structures once they were subjected to a particular combination of loads. We had to make guesses as to points of inflection, maximum bending, and other physical results of the loads.
>
> We were required to keep comprehensive notes of all lectures. At the end of each semester those notes had to be handed in, and were graded. He insisted that we put our name and the date on every piece of paper we touch, before commencing any work or notes. I continue that practice to today.
>
> Having graduated as class valedictorian from a liberal arts college at age 17, Professor Cross was unbelievably familiar with the great authors and important contributors to world literature. He often sprinkled his lectures with allusions to literature that had no direct or obvious connection to engineering or mathematics. One of his favorite such resources was Lewis Carroll's *Alice in Wonderland*. However, there were innumerable others. Being hard of hearing, Professor Cross had a hearing aid as long as I can remember. The controls were located in his vest pocket, and he was constantly adjusting the volume—up or down—depending upon the needs of the moment. At times, I must say we wondered whether he sometimes chose to "tune out" when conversation became too mundane. He also was very emphatic about the way we asked questions. His position was that it is generally more difficult to ask the right question, than it is to get the right answer. He believed that great engineers were those who had learned how to ask the right questions.[17]

As a clue to Cross's own methods of teaching, we may turn to the notes of Thomas Kuesel. While Cross was department chair at Yale, he continued to teach. His course was Civil Engineering 102 (Structural Engineering), and students came from all over the world to take it. By great good fortune two notebooks kept by Thomas R. Kuesel have survived. Kuesel was an able student who later became one of Cross's assistants. Cross used a red pencil to correct Kuesel's notes, which are mostly in ink. On the basis of internal evidence, we believe that the course was set up for beginning graduate students. While offering some useful definitions, Cross does not take up such fundamentals as Hooke's law. These would have been covered in an earlier course.

Kuesel's notes are contained in two black-bound volumes. The first was from March 4, 1946 to June 18, 1946. The second runs from September 16, 1946 to May 28, 1947. All are on graph paper in an excellent hand. There are a great many diagrams and tables. Kuesel was a capable engineering draftsman. These notebooks unquestionably provide a remarkably clear exposition of the mature thought of Cross during the years when he was at Yale (1937–51).

The difficulty with using material of this type, as the historian will quickly point out, is the problem of knowing what is Cross and what is Kuesel. Fortunately Kuesel had the useful habit of boxing or underlining any statements which seemed especially important. As a description for this material I have decided on Kuesel, *Notes on Cross,* I or II, with a following page number as the appropriate citation. Kuesel numbered his pages with remarkable, but not absolute, consistency. The table of contents in volume one indicates that Cross took up, in order, the following subjects:

(1) *The Field of Structural Engineering*
(2) *Analysis of Stresses and the Interpretation of the Analysis*
(3) *Deflected Structures*
(4) *Moment Distribution*
    Informal Moment Distribution
(5) *Reinforced Concrete*
    * Standard Textbook Procedure for Determining Stresses
    * Design of Concrete Beams and Girders
    * Books on Reinforced Concrete
    * Concrete Columns in Flexure
    * Four Ways of Arriving at Truth in Structure
    * The Textbook Theory of the Action of Reinforced Concrete Beams
    * Specifications
(6) *Pressure Lines*
    * The Theory of Strings
    * Fit and Balance
    * Deflected Structures Applied to Arches
    * Fixed End Beams
    * Horizontal Loads
    * Formal Procedure—the Column Analogy
(7) *The Effect of Deformations in Structures*
    * Temperature Stresses and Shrinkage
    * Rib Shortening

* Rotation and Spreading of Abutments
   * Settlement of Supports in Continuous Beams
(8) *Influence Lines*
   * Application to Frames and to Arches
   * Stresses in an Arch
   * Some Notes on Deflections
   * Arch Dams and Waterbanks
(9) *Prestressed Concrete*—Guest Lecture by Professor Magnel

Gustave Paul Robert Magnel (1889–1955) was a Belgian engineer who made major contributions to the theory and practice of pre-stressing and post-tensioning reinforced concrete. Having graduated from the University of Ghent in 1912, Magnel moved to London in 1914. He spent the next four years in that city, where he worked with the Somerville firm. By 1917 he was named chief engineer for Somerville and sent to Paris to study French methods of reinforced concrete construction. In 1919 he moved back to Ghent where he began his university teaching career. In 1922 he was authorized to develop a course on the theory of reinforced concrete, and in 1930 he established a testing laboratory. In the course of time, Magnel found a system of pre-stressing which was an improvement on the method of Eugene Freyssinet, the first engineer to make pre-stressing possible. Freyssinet, however, used mild steel of low tensile strength. Magnel's innovation was to use steel wire of high tensile quality. He showed that steel could be stressed up to high values without reaching its elastic limits. His innovations quickly became accepted practice.

Magnel also designed a number of small bridges and industrial structures in Belgium and the Netherlands. Some of the bridges show considerable elegance. These include: the Railroad Bridge at the Rue Des Miroirs, Brussels; the Sclayn Foot Bridge, Namur; the De Smet Bridge, Ghent.

Illustration of Magnel's work are available in his book, *Prestressed Concrete: Its Principles and Applications* (New York, 1954). Pre-stressing or post-tensioning techniques began to be used in the United States only in the late 1940s and early 1950s. At this writing (2002) these practices have been common for many years.

The table of contents in Kuesel, *Notes on Cross,* II, is as follows:

1. Definition of Terms
2. The Field of Structural Engineering
3. Spandrel-braced Steel Arches
   * General Consideration
   * Layout
   * Stresses in a Steel Arch

* Live Load Stresses
* Temperature and Wind Stresses
4. Continuous Trusses
5. Temperature Stresses in a Steel Arch (con't)
6. Book Review Section
7. Precise Analysis in Indeterminate Structures
8. The Displacement Diagram
9. Quiz Section
   * Wind Stresses
   * Vibrations
   * Earthquakes
   * "Cost" of Investigation (Evaluation)
   * Wind Stresses in High Building—a Comparison of Methods of Analysis of Buckling
   * A Study of Buckling in Steel and Aluminum Columns
   * Factor of Safety
   * Plasticity
   * Predesign of Indeterminate Structures
   * Secondary Stresses "Participation"
   * Column Action
   * Participation Stresses in a Bent
   * Predesign of a Two-Hinged Frame
   * Suspension Bridges
   * Bracing
   * Slabs[18]

In the Kuesel *Notes,* there are many references to contemporary works of engineering, particularly the Golden Gate Bridge, the Tacoma Narrows Bridge, and the Ambassador Bridge. There is much less reference to contemporary architecture, though Cross was very clear that most reinforced concrete beams would be designed to fit the specifications of an architect. There are no references to structures in other countries or to historic structures.

In the two volumes of notes there is only one statement surrounded by quotation marks. It is slipped into the opening page of Volume II and is as follows: "Many times in civilization men have asked the wrong questions. They have answered the question with accuracy. But they have failed to recognize that they had the wrong question" (Hardy Cross, May 6, 1947).

Cross was obviously trying to educate engineers who would ask the right questions. We proceed now to a few remarks which will, we hope, give an idea of the richness of those notes.

From all accounts Cross thought of himself, justifiably, as an expert in many aspects of civil engineering. In the following pages we discuss those sections of the Kuesel notes which are boxed or underlined. We think that they are a good reflection of the mature thought of Hardy Cross. The reader will find that the passages which we have chosen are, to say the least, unvarnished. In fact, part of the immense usefulness of these notes lies in the manner in which they reveal the undisguised views of the teacher. These are often opinions which Cross did not care to publish. A good example is his comment on specifications which Kuesel both boxed and underlined: "Specs are not and should not be completely logical. They are a mixture of theory, experience, the results of tests, plus frequently some commercial pressure." A little further down the page we find: "The important thing is to know what is worth figuring, not how to figure it."[19]

One has the feeling that Cross was trying to convey a kind of engineering wisdom. His journal papers are well worth reading even today but his classes must have been a rare experience.

Cross was a man who defined his terms carefully, and he was especially careful in dealing with words whose meanings were in dispute. "Stiffness," for example, was a word which was obviously important, but the profession had " ... only a vague idea of what it is."[20] Cross was fond of irony. Early on he used a definition from Webster: "*Epistemology:* the theory or science of the method and grounds of knowledge, especially with reference to its limits and validity." In his thought he was intent on separating observed fact from shrewd guesswork and/or faulty deduction. In any discussion he liked to split off those aspects of a problem which were inarguable from those aspects which might properly be taken up. The laws of statics were an example of the former. The properties of materials, especially those not fully investigated, were an example of the latter.

This attitude led Cross to dispute a good many beliefs of his contemporaries in a manner which could be both humorous and biting. Consider a statement on the question of the bonding action between concrete and its reinforcements in a beam. At the top of Kuesel's page is a cow before which three figures are genuflecting. A formula follows with explanatory terms. Finally there is an injunction to the student which encapsulates the entire problem. The bonding action between concrete and its reinforcements is better understood today. Even so, it would be hard to think of a better illustration of the situation (Figure 16), p. 20. Kuesel, *Notes,* I.

Another example of a Hardy Cross put down is his characterization of the textbook theory of the action of reinforced concrete beams. "This," he said, "is perhaps the most useless mess of nonsense ever concocted by the human brain—however, it usually works."[21] One concludes that a good student who

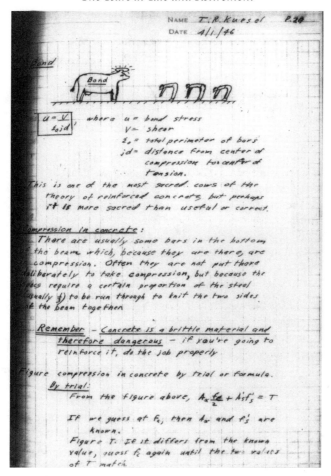

Figure 16. Bonding action between concrete and steel. Kuesel, *Notes*, I, p. 20.

worked hard was bound to emerge from this course with a solid knowledge of structural mechanics (as it then existed) and a hearty skepticism of much generally received doctrine.

Cross's methods inevitably led him to comment on the history of structural engineering and on the literature of his field. At the beginning of the century, he said, the most advanced theoretical work was going on in France. America had done an immense amount of building (he was undoubtedly thinking of the great transcontinental railroads) but had produced no important body of theory. Then the United States started to send men abroad, to France and increasingly to Germany. Now (1946) we could stand on our own and need not rely on European specialists. Cross was modest enough to pass over his own contribution. For anyone interested in the history of reinforced concrete, the bibliographies of Hardy Cross and his comments are fascinating documents. On April 18, 1946, he offered a bibliography on reinforced concrete. The list reflects not only his wide acquaintance but also his frank opinions:

> Turneaure & Maurer 1907, 4 or 5 editions.
>
> Turneaure was the dean of engineering at the University of Wisconsin—retired some years ago—one of the most discriminating investigators in the field: He wrote books on reinforced concrete, water supply, and a series on the theory of structures. *Sutherland and Reese* (Now out of fashion but still useful.) Sutherland was head of civil engineering at Lehigh; Reese a consulting engineer. This is a revision of Sutherland and Clifford, brought into line with modern practice. *Taylor, Thompson, and Smulski.* One of the earlier books. Taylor had nothing to do with it, and Smulski was a student of Thompson. Voluminous but highbrow—not very useful.
>
> *Urquhart and O'Rourke.* A summary for classroom purposes—not very good.
>
> *Hool and Othert.* Hool was at the University of Wisconsin at Milwaukee and wrote a series of books on concrete and other phases of structural engineering. It is doubtful whether he was ever an authority on anything—he had a genius for editing things and getting them out. He was not personally well known. There is a lot of material if you care for it—much out of date.

There is much other bibliographical information in Kuesel's *Notes,* but this short list is indicative. We conclude that Hardy Cross was never diffident about stating his view on the quality of work in the profession of civil engineering, both inside the academy and in the world of practical affairs. One suspects that he was equally willing to give his opinion in a variety of other fields.

A particularly notable aspect of the *Notes* is the emphasis which Cross placed on civil engineering as a high moral endeavor. He defined civil engineering as "planning for the use of land and air and for the control and use of water and designing, constructing, and operating the works needed for the carrying out of this plan."[22] Today it would be difficult to find a textbook which stressed this side of the profession. Most authors conceive a potential client with funds to afford the most up-to-date building technology. It is up to them to supply it. Hopefully the growing side of civil engineering called "failure analysis" is a reaction to this state of affairs. In an excellent chapter on "The Worst Structural Disaster in the U.S.," Mario Salvadori faults the design engineers of the Hyatt Regency Hotel in Kansas City for accepting a dangerous suggestion by the contractor for simplifying the walkways. Hardy Cross would have considered their failure to be a mortal sin. The Kuesel *Notes* are full of injunctions such as "*Remember—Concrete is a brittle material and therefore dangerous—if you're going to reinforce it, do the job properly.*"

"In teaching," Cross said, "we must work backward. Take up analysis first to obtain facility in general and later in more exact methods of stress analysis. Once these have been learned, we may turn to interpretation. Then to design—'structural morphology' forms and types of structures and purposes to which they may be adapted."[23] In keeping with this concept, Cross began his course with some fairly simple analyses of deflection in beams and girders. He introduces the idea of moment distribution early on, and of course, insists on the conventional signs for indications of moments and shears. He has a provocative section on the possible sources of error in the procedure (difficulty with the calculation of fixed-end moments was the most common). There follows a long section on reinforced concrete which contains the observation that in order to compete with steel reinforced concrete had to be "scientific" and that its action was not understood for a considerable period.[24]

Cross deals with concrete columns in flexure, with longitudinal stress in concrete, and with various other problems in the material. Obviously it was his favorite. Arches receive a substantial treatment, and there is a fine diagram (Figure 17) with a suggestion that may refer to Nathan Newmark's article, "Interaction of Rib and Superstructure in Concrete Arch Bridges," *Transactions* A.S.C.E. 103 (1938), pp. 62–80. (Not the *Engineering News Record,* as Kuesel has it.)

This drawing is of great interest since it throws light on one of the puzzles in Hardy Cross's career: his failure to comment on the works of the early European masters of reinforced concrete, notably Maillart, Nervi, and Torroja. We have noted that Mr. and Mrs. Cross traveled extensively in Europe during the 1930s and that they did not call at the office of Maillart in Switzerland or meet Pier Luigi Nervi in Italy.

At this point personal considerations enter the story. In his mature years, Cross became quite deaf. At Yale he always wore a hearing aid in the classroom. Neither he nor Mrs. Cross spoke a foreign language. Maillart spoke no English, and at that point in his career, neither did Nervi. A personal encounter would therefore have been difficult. The drawing shows that Cross was interested in the deck stiffened arch. It is easy to see the lowest sections as a kind of crude version of Maillart's magnificent Salginatobel Bridge of 1930 (see Figure 7 in Chapter 2). Cross, however, shared Newmark's reservation that, "there is no readily available and convenient means of analyzing this as a unit." To Cross, and to many of his American contemporaries,

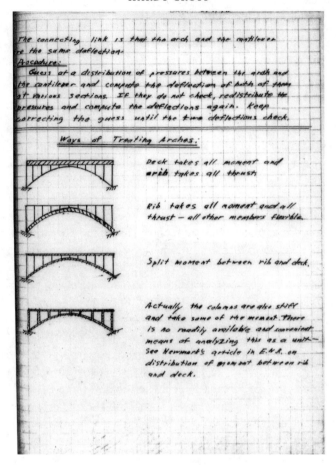

Figure 17. Ways of treating arches. Kuesel, *Notes,* II, p. 25.

Maillart must have seemed like a magician. They had to grant the supreme elegance of his structures, but could not say how it was achieved.

The second volume of notes begins with recapitulation of terms (for the new semester) and proceeds quickly to topics in which Cross was interested. These matters included bridges of various types. Hence he paid attention to wind-bracing, to many types of truss, and to a variety of other questions. It is well to observe that by 1946 there had been a good many failures of large bridges and that Hardy Cross had evidently studied them all. In *Objections to Continuous Beams* he showed:

There is no way of erecting a continuous rolled beam so that you know what you're doing. Are the supports on the same level? How accurately can you measure them?

You can insert a jack and jack with a known force, to determine the stresses under some condition of loading (e.g., full dead load). If you overjack, you can cause the end reactions to be so near that live load may cause the ends to lift off their supports.

The settlement of supports is very difficult to predict. We can however, discuss a settlement of such a magnitude as to put the bridge under suspicion.[25]

The portion of the notes on "bridges" has many such passages.

Another book review section of seven pages (37–43) in *Notes* II, contains Cross's reflections on the development of his field as well as comments on his contemporaries and their treatises. He begins with Maxwell's theorem and remarks that the Germans took it up and elaborated it mathematically. In a certain sense, Cross wanted to take most of the mystery and a lot of the mathematics out of civil engineering. Hence he held that the influence of German mathematics on structural theory was pernicious. He wrote that people came back from Germany and wrote books which " . . . had defects in that they told you practically nothing about how to get the structure you were going to analyze and have no discussion of the fact that whether or not you could solve the equations depended on when and where you cut."[26] The bibliography is too lengthy to quote here, but we must note the curious attitude of Cross toward Timoshenko. Cross listed all of the latter's publications and, concerning Timoshenko's *Theory of Structures,* remarked, "another point of view—distinctly European."[27] For Hardy Cross this comment probably meant "overly mathematical." We are led to reflect on the apparent antipathy between the two giants of structural mechanics at mid-century and to suggest that it may have had its source in education at the secondary school level. Timoshenko enjoyed the advantages of a mathematical education in a European gymnasium. These were, one suspects, substantially greater than those of Norfolk Academy. Timoshenko saw advanced mathematics as a fundamental requisite for structural design. Cross, on the other hand, was skeptical of elaborate mathematical solutions. The advent of the computer has obscured the conflict between the two men, but it was a very real difference for over two decades. About Cross and Morgan, *Continuous Frames of Reinforced Concrete,* the primary author simply remarked "some good ideas, many need revisions."

To earthquakes Cross devoted only a few pages in comparison with the attention given to bridges, especially suspension bridges. Cross was evidently fascinated by this form. He had, after all, been on the committee which reported on the failure at Tacoma Narrows a few years previously. His comments on the heat treatment of wire at the Mt. Hope and Ambassador Bridges are full of questioning remarks. "These bridges," he noted, "are very temperamental—they will suddenly go on a rampage for no clearly discernible reason and then settle down for years."[28] He also notes the difficulty of obtaining knowledge from an article by an expert in one of the trade journals. Too much money is involved.

Finally there is a long discussion of flat slabs with many references to the history of the slab and to C. A. P. Turner, whom Cross felt had befogged the issue with terminology. In the early patented systems the difference was clearly between two-way and four-way reinforcement. Kuesel illustrated the Smulski system, which, he noted, was difficult to erect but produced an excellent bond. There was almost no lapping of reinforcement. In this section of the notes, one has the impression that the instructor was clearly not happy with the research on the flat slab up to 1946. Apropos of a series of articles, Kuesel boxed the statement, "The important thing is that you can't make head or tail of the data."[29] And one is tempted to ask, "If Hardy Cross couldn't interpret the data, who could?" Here, as with Maillart's bridges, it seems clear that Cross was unfamiliar with the Swiss designer's improved flat slab, which was published in the *Schweizerische Bauzeitung* (22 May 1926), pp. 263–65. On the whole, we believe that the Kuesel *Notes* are the best evidence we have on Cross's thought in his mature years. Unfortunately there is no equivalent for his period at Urbana-Champaign. We can turn to the pages of *Engineers and Ivory Towers* for his views on matters outside the classroom.

It is clear that Cross wanted his students to have a broad liberal arts background such as he and his brother had received at Hampden-Sydney. He fulminated against over-specialization in the engineering sciences. Cross often remarked that civil engineering is a cultural subject and quoted with approval a statement by James Bryant Conant, President of Harvard University, that the separation of the branches of learning had gone too far.[30] For Hardy Cross engineering was an art, an art in the service of humanity. He liked Thomas Tredgold's definition: "Engineering is the art of directing the great sources of power in nature for the use and convenience of mankind" and used it as the introduction to the last essay in *Engineers and Ivory Towers*. In all probability this piece was an address to the Institution of Civil Engineers of Great Britain, whose Gold Medal he received in 1958. (Cross was the only American ever so honored.) In his speech, Cross sketched the progress of America in the history of its canals and railroads, and with great confidence declared that in the future it would not be necessary for Americans to turn to Europe for technical thought. With appropriate deference to his hosts, he used a long quotation from Robert Louis Stevenson, the son and grandson of engineers. Cross warned that poor engineering meant failure and misfortune, inconvenience, suffering, and death, and recommended that anyone who believed that material progress meant moral decay read Daniel Defoe's *Journal of the Plague Year*. Long before C. P. Snow announced "The Two Cultures," Cross was a man who bridged them perfectly (no pun intended).

There was probably not a "humanist" on the Yale Campus with whom Hardy Cross could not have conversed intelligently. Of course he would have bothered a good many people with seemingly innocent fundamental questions about their disciplines. It is a great shame that his deafness probably kept him from enjoying the social interaction with his colleagues which he valued so greatly.

Personally Cross seems to have been approachable but somewhat reserved. Probably because of his deafness, he did not make friends easily. He was devoted to his wife, the former Edythe Fenner, but she was rather frail. She accompanied him to Hampden-Sydney when he received an honorary degree from his alma mater in 1935, and those who met her described her as "most attractive." She died in 1956, and is buried with him in the Wright family plot at Ivy Hill Cemetery near Smithfield, Virginia. They had no children.

In any large institution there are inevitably a number of committees. As a former professor in a large research university, I recognize in Hardy Cross a man who would have been an ideal member of many of these committees. I can recall sitting on one Fulbright interview board where most of the award candidates were musicians or musicologists. My own field was architectural history, but someone probably put me on that board because I was known to have come from a musical family. I think I made a contribution. At any rate, I cannot imagine any interdisciplinary committee in the sciences or the humanities where Hardy Cross would not have been an invaluable member.

At the same time I can see that justice of Robert Goodpasture's remark to me that, "Yale never knew quite what to do with Hardy Cross." There are clear reasons for this. He could be stubborn. He could be dogmatic. It is suggestive that he never revised his classic textbook of 1932, written with Newell D. Morgan, *Continuous Frames of Reinforced Concrete*. Its fourteenth printing was in 1954. It is inconceivable that there were no new additions to the subject in between 1932 and 1954. What, indeed, should a university do with a personality such as Hardy Cross?

One answer is to leave him alone. Another is to call on him, most diplomatically, for service to the entire academic community. This, Yale apparently did not do. A distinguished acquaintance, whose years at Yale overlapped those of Cross (1937–51) remarked to me that, "In those days the engineering school was somewhat off by itself."[31] My friend knew one or two of the civil engineers, who had joint appointments in architecture, but he did not know Cross. One has a certain sympathy with the Yale administration. They had brought in a somewhat assertive genius whose fame was worldwide (students came from all over the globe to study with Cross), but there was no doubt that he was a difficult member of the university community—which he nevertheless esteemed highly. Yale's solution was to allow Cross to develop his own private fiefdom in the department of civil engineering. In that department, he built a strong faculty which was devoted to him personally, and his students were as enthusiastic as ever.

When he retired in 1951, he received the customary tributes from his colleagues and from the Yale administration. But during his years at Yale, aside from *Engineers and Ivory Towers*, we find no publications of any importance. Hardy Cross was an effective department chairman and, as always, a superlative teacher. Those two tasks absorbed all his energies. His most important work was done at Illinois.

Cross served under two deans at Yale: Samuel W. Dudley (1936–48) and Walter J. Wohlenberg (1948–55). Dudley was a graduate of the Sheffield Scientific School and a former chief engineer for the Westinghouse Air Brake Company. Well-versed in several phases of engineering, Dudley evidently got along well with President Charles Seymour, (B.A. 1904, M.A. 1909, Cambridge; B.A. 1908, Ph.D. 1911 Yale). Seymour was, says a historian in *Engineering at Yale*, "the quintessential Yale man." He had been Sterling Professor of History (1922–37) and provost (1927–37). He is described as: "a quiet, thoughtful, conservative person, not given to precipitate action."[32] Despite the difficulties of the war years, the engineering school evidently did reasonably well under the Seymour presidency. The outlook changed perceptibly when the corporation appointed A. Whitney Griswold to take over the helm of the university in 1950. Griswold was a strong believer in the liberal arts, and, says W. Jack Cunningham, "There were those in engineering who felt that Griswold had no interest in the sciences or engineering, even feeling that he might want to do away with engineering at Yale.[33]

These fears were intensified when Griswold, shortly after assuming office, chose to make inquiries about the engineering school not through Dean Wohlenberg but through Hardy Cross. Apparently Griswold was impressed by the fact that Cross was the only man on the engineering faculty who had began his studies with a liberal arts degree. "Evidently," says Cunningham, "Griswold had more trust in information from a liberal arts graduate than from an engineering graduate."[34] One would like to hear a record of that meeting: Cross, the nineteenth-century Southern gentleman, sure of himself after a lifetime's accomplishment and recognition all over the world, and Griswold, the newly minted president of an important Ivy League University. From his remarks at the Yale Club in New York City on November 13, 1952, we suspect that Cross was not overly impressed with Griswold. In any event, his impression at the interview would have been of little importance. Cunningham remarks that, as his presidency progressed, Griswold initiated a series of investigation into the engineering school which culminated in a major upheaval that started in 1960 and continued for more than a decade. A. Whitney Griswold was a great president for Yale University. Whether he was so good for the engineering school is open to question.

Among Cross's papers at Urbana-Champaign is a complete transcript of a talk given by him on November 13, 1952, at a meeting of the Yale Engineering Association at the Yale Club, New York City. It is remarkably frank. He characterizes his two years at M.I.T. as "vocational training." It is also clear that he had no high regard for President Griswold. He objected to Griswold's use of the word "aesthetics." He was also skeptical of Yale's emphasis on research and development, though he exempted from his strictures the very distinguished work done in metallurgical engineering by Dr. Mathewson. Cross did not believe that the concept of graduate work had been clearly thought out in America in any field, and he specifically mentioned philology despite the fact that his own brother was a philologist. He closed with a poem which he himself had written:

> Teach too much absurd precision
> Scam the common human vision
> Seek entirely views pedantic
> Use too many words semantic
> So amuse as heads roll by
> Both the students and alumni.

Having noted this career pattern, we must ask: Was Hardy Cross wise to leave Urbana and move to New Haven? Would he have continued his brilliant research career if he had stayed at Illinois? To this question there can, of course, be no real answer. In 1936 he had published his important "Analysis of Flow in Networks of Conduits or Conductors." This paper marked the application of his methods of geometry and of successive corrections to an entirely new kind of problem. If he had continued his investigations, he might have made additional major contributions to hydraulic and electrical engineering. Or he might have run into a dead end. He would certainly have been lonely after Harald Westergaard moved to Harvard in 1936. And

Hardy Cross in Britain receiving the Gold Medal of the Institution of Civil Engineers. It was the first to be awarded to an American.

the retirement of Wilbur Wilson was in sight. In all probability his wife was happy to be closer to her family in Rhode Island, so the move to Yale was probably a good one.

When the Crosses left New Haven, they moved to a cottage at Virginia Beach where they had a housekeeper/nurse for Mrs. Cross. Her name was Inez Borthwick. We have noted earlier that Cross was always comfortable in Tidewater, Virginia, and the couple undoubtedly enjoyed the mild winters there. From all accounts they had a quiet retirement. Hardy Cross enjoyed fishing. There are photographs of him with some rather large tarpon. He also continued to read a great deal. Three of the books which were with him at his death on February 11, 1959, were paperback editions of Gordon Home's *Roman Britain*, Conan Doyle's *The Sign of the Four*, and William Dean Howell's *Venetian Life*.

The great event of these last years was undoubtedly his trip to Britain to receive the gold medal of the Institution of Civil Engineers. (At the London headquarters of the Institution of Structural Engineers on Upper Belgrave St. there is today a Hardy Cross room.) On this trip he apparently paid close attention to the network of canals which were a legacy of British progress in the early phases of the industrial revolution. He made a special trip to look over the Manchester Canal and was a firm believer in the feasibility of a canal through the Dismal Swamp. It would be coupled with a bridge fourteen or fifteen miles long crossing Chesapeake Bay. Today the bridge in which he was so interested is a reality. It is part of the great bridge-tunnel system linking the western shore to the mainland, from Norfolk to Virginia Beach. However, the canal through the Dismal Swamp has not been built. Shortly before his death, Cross was notified that he was to receive the gold medal of the Franklin Institute in Philadelphia. It was his last formal award.

In our time Hardy Cross is remembered with an annual competition among engineering students in Virginia. But his real memorial is the thousands of buildings around the United States designed with a moment frame. The previously mentioned Equitable Building by Pietro Belluschi and Mies van Der Rohe's Seagram Building are only two of the outstanding high-rise buildings with this kind of frame. It was Hardy Cross who discovered a structural system, which eliminated the interior shear wall and allowed an immense saving of time and material. The new crop of postwar high-rise buildings in the United States were not only more elegant solutions than the older work, they were also less costly and more efficient. This circumstance was largely the result of the application of Hardy Cross's theories. The advent of the computer should not obscure his contribution to engineering and to architecture. Surely he belongs in the company of James Clerk Maxwell and Otto Mohr. His compatriots owe him more recognition than he has so far received.

Appendix A: Glossary

Since this book has been written to introduce Hardy Cross to a non-professional audience, I have included a glossary of technical terms. To put it together has been no easy task. Structural engineering is partly based on physics and partly on mathematics. Furthermore a strong visual imagination is required.

Beam—is the horizontal structural element which connects the tops of two columns. It is subject to bending. If it is thickened at the ends, it is said to be "haunched." A beam rests on an upright structure called a post or column. The beam may or may not be continuous. The exterior colonnade of the Parthenon is a perfect example of a post and beam construction. It is not, however, continuous. In a frame building the columns and beams are multiplied so that if there were no walls or partitions the effect would be that of a jungle gym.

Bent—a framework transverse to the length of a structure (as a trestle, bridge, or long shed) usually designed to carry lateral as well as vertical loads. *Webster's Third New International Dictionary.*

Carry-over Moments—the distributed moments in the ends of the members cause moments in the other ends, which are assumed fixed, and these are the carry-over moments. From Jack C. McCormac, *Structural Analysis* (Scranton, Pa., 1962), 330–31.

Continuity—refers to the transference of structural action. Cross notes that a concrete beam cannot bend without deforming the girders and columns connected to it. He also observes that it is more important to recognize this fact than it is to evaluate it precisely.

Distributed Moments—after the clamp at a joint is released, the unbalanced moment causes the joint to rotate. The rotation twists the ends of the members at the joint and changes their moments. In other words, rotation at the joint is resisted by the members and resisting moments are built up in the members as they are twisted.

Rotation continues until equilibrium is reached—when the resisting moments equal the unbalanced moment—at which time the sum of the moments at the joint is equal to zero.

Fixed End Moment—is a moment generated by a beam at an end which is bolted (steel) or otherwise firmly fixed in place. In concrete construction the column and beam are poured simultaneously so that the building is monolithic.

Flexure—the deformation of an elastic body wherein all points originating in a straight line are displaced in the same plane to form a curve.

Frame—as used by Cross, this word generally refers to the skeleton of a building. In the typical modern skyscraper this frame consists of columns, beams, and girders. The Empire State Building is a good example.

Hinge—a joint which allows unrestrained rotation. In reinforced beams columns the hinge is usually provided by crossing the reinforcing base in order to reduce the bending capacity of the section.

Hooke's Law—the stress within an elastic solid to the elastic limit is proportional to the strain responsible for it. Mathematically expressed as s/s.

Modulus of Elasticity—the ratio of stress in a body to the corresponding strain. This ratio will vary in different materials. It will be different in wood, steel, and reinforced concrete. In the last material it may vary with the placement of the reinforcing bars.

Moment—refers to the action of a beam under load. Since the beam itself is partly in compression and partly in tension, the moment is always present whether or not it is visible. It is a measurable quantity and is expressed in foot pounds. Cross thought of it as "the algebraic sum of the products of force times distance on one side of a section about an axis lying in the section."

Moment of Inertia—a quantity which shows the bending stiffness of a cross sectional shape, usually a beam. The moment of inertia is proportional to the area of the cross section, i.e., to the amount of material used, but also, and more significantly, to the square of a length called its radius of gyration, that measures the moving of the material away from the neutral axis. The deflections and bending stresses in a beam are proportional to the moment of inertia of its cross section. Moments of inertia appear in standard section tables, so that the designer can obtain the strength and rigidity of a given beam without lengthy calculations. Salvadori and Heller, *Structure in Architecture: the Building of Buildings*, (N.Y., 1986, 3rd ed.), 144.

Portal—any vertical space between two uprights included between two horizontals (as of floor and ceiling) which must be kept open for free communication in a building of skeleton construction.

Shear—is the state of stress in which the particles of the material slide relative to each other. Rivets in riveted connections tend to shear. The weight of a short cantilever beam built into a wall tends to shear off the beam at its root.

Shear introduces deformations capable of changing the shape of a rectangular element into a skewed parallelogram. The shear strain is measured by the skew angle of the deformed rectangle rather than by a unit length, as in the case of

Appendix A

tension or compression. Salvadori and Heller, *Structure in Architecture*, (3rd ed., N.Y., 1986), 94.

Slab—Cross here refers to the use of reinforced concrete as a slab supported by a column as opposed to the employment of the material in a frame. In the United States this system was popularized by the Minneapolis engineer, C.A.P. Turner, who attempted, unsuccessfully, to patent his method in 1908. Almost simultaneously Robert Maillart developed a more elegant system in Switzerland. For illustrations of the Turner system, see Leonard K. Eaton "Oscar Eckerman: Architect to Deere & Co., 1897–1942," 113–29 in *Gateway Cities and Other Essays* (Ames, Iowa, 1989).

Stiffness—as used by Cross, this term means " . . . the moment which would exist at the ends of the member if its ends were fixed against rotation." *Arches*, 1.

Structural Redundancy—essentially allows loads to be carried in more than one way, i.e., through more than one path through the structure. It is a needed characteristic in any large structure or any structure whose failure may cause extensive damage or loss of life.

The *Three Equations of Statics*—are simply statements of Sir Isaac Newton's observation that for every action on a body at rest there is an equal and opposite reaction. Newton stated that a body at rest is in equilibrium. The resultant of the external loads on the body and the supporting forces or reactions is zero. The sum of the horizontal forces is zero. The sum of the vertical forces is zero. And the sum of the moments in of these forces about any point in the plane is also zero.

Unbalanced Moments—initially the joints in a structure are considered to be clamped. When a joint is released, it rotates because the sum of the fixed moments at the joint is not zero. The sum of the member-end moments at an initially clamped joint that violates joint equilibrium in the unbalanced moment.

Appendix B:
Selected and Annotated Bibliography of the Writings of Hardy Cross

In the Journal of the American Concrete Institute

1. "Continuity as a Factor in Reinforced Concrete Design." *Proceedings* 25 (1929).

   Method of distributing moments: computation of maximum moments, shears and reactions including effect of haunching, columns, joints. Economy.

2. "Design of Reinforced Concrete Columns Subject to Flexure." *Journal* 1 & 2 (Dec. 1929): 157–69.

   Variation of design procedure depending upon whether the column moment varies much with respect to its stiffness, or not; whether the moment is due to deformation or load. Need of factor of safety relative to tensile steel. Use of flared column heads.

   Discussions
   W. M. Dunagan. *Journal* 1, no. 6 (Apr. 1930): 656–59.
   Stanley G. Cutler. *Journal* 1, no. 6 (Apr. 1930): 656–59.
   George H. Dell. *Journal* 1, no. 6 (Apr. 1930): 661–65.
   N. H. Roy & P. E. Richart. *Journal* 1, no. 7 (May 1930): 775–92.
   "Closure." *Journal* 1, no. 7 (May 1930): 792–98.

3. "Simplified Rigid Frame Design." *Journal* 1, no. 2 (Dec. 1929): 170–83.

   The "Moment Distribution" method of analysis as a simplified procedure for use in ordinary cases of rigid frames. A straightforward explanation of this method and the value of its application.

   Discussions
   Frank J. McCormick. *Journal* 1, no. 6 (Apr. 1930): 666–68.
   T. F. Hickerson. *Journal* 1, no. 6 (Apr. 1930): 668–69

4. "Why Continuous Frames?" *Journal* 6, no. 4 (Mar.-Apr. 1935): 358–67.

Economy and general considerations of a rigid frame bent as a continuous frame. Idea of predesign in rigid frame buildings.

Discussions
Arthur G. Hayden. *Journal* 7, no. 1 (Sept.-Oct. 1935): 121–22.

5. "Rigid Frame Bridges" (synopsis). In "Report of Committee 314—Hardy Cross, Chairman." *Journal* 10, no. 2 (Nov. 1938): 69–71.

Purpose of Rigid Frame Bridge, approximate stress distribution, importance and interrelation of assumptions of material and its properties.

Discussions
Charles S. Whitney. *Journal* 10, no. 5 (Apr. 1939): 72–81.
"Closure." *Journal* 10, no. 5 (Apr. 1939): 72–82.

In the Transactions and Proceedings of the American Society of Civil Engineers

1. "The River and Harbor Problems of the Lower Mississippi." *Transactions* 87 (1924): 1147–50.

A National Hydraulic Laboratory would be of practical value to the nation as a clearinghouse for standardization of terms, and would qualitatively improve study of river problems and collection of field data.

Discussion of Paper #1545 A Symposium. *Transactions* 87 (1924): 972–1097.

2. "Concrete Specifications." *Proceedings* (September 1925): 1526–37.

Discussion of variations from the First Committee Report likely to affect standard practice in design, particularly criticizing "over-formularization," too strict rules for computation of bending moments, unclear treatment of flat slabs and footings, and general disrespect of clear bases for recommendations.
Discussion of "Standard Specs for Concrete and Reinforced Concrete." *Proceedings* (October 1924): 1153–1285.

3. "Design of Symmetrical Concrete Arches." *Transactions* 88 (1925): 931–1029.

Great precision in arch computations is not justified. Uncertainties concerning settlement, loadings, temperature stresses make great refinement absurd.
Discussion of Paper #1568 by Charles S. Whitney. *Transactions* 88 (1925): 931–1029.

4. "Design of a Multiple Arch System." *Transactions* 88 (1925): 1142–82.

An algebraic analysis of an arch system provides more accurate results, is more flexible and considerably less expensive and complicated tool than an analysis by the theory of the ellipse of elasticity.
Discussion of Paper #1571 by A. C. Janni. *Transactions* 88 (1925): 1142–82.

5. "Secondary Stresses in Bridges." *Transactions* 89 (1926): 1–149.

Suggestions for simplification of several methods used in determining secondary stresses. Reminder of uncertainties always entering into the problem.

6. "Flood Flow Characteristics." *Transactions* 89 (1926): 1047–48.

The design of river crossings is important and scientific information for this is not as available as for the actual structural design of bridges. A big problem is provision for flood flow and resultant erosion.
Discussion of Paper #1589 by C. S. Jarvis. *Transactions* 89 (1926): 985–1032.

7. "Moments in Restrained and Continuous Beams by Conjugate Points." *Transactions* 90 (1927): 1–49.

This method "really treats of geometry" and consequently the relations can be presented in several different ways mentioned.
Discussion of Paper #1598 by L. H. Nishkian and D. B. Steinman. *Transactions* 90 (1927) 1–49.

8. "Virtual Work: A Restatement" Paper #1605. *Transactions* 90 (1927): 610–18.

Extension of Virtual Work to determination of deflections and rotations in beams, deflections and angle changes in trusses. Note that geometrically there is no question about the results. However, this approach doesn't consider the validity of the physical assumptions regarding the action of structures.

Discussions
E. A. MacLean. *Transactions* 90 (1927): 619.
Richard G. Doerfling. *Transactions* 90 (1927): 619–23.
J. Charles Rathbum. *Transactions* 90 (1927): 625.

Closure
General consideration of criticisms contained in discussions listed above.
Transactions 90 (1927): 623–26.

9. "Analysis of Continuous Concrete Arch Systems." *Transactions* 90 (1927): 1137–45.

Appropriate and more exact methods of analysis of continuous arch systems. Consideration of foundation distortions and influence of the floor system.
Discussion of Paper #1621 by Charles S. Whitney. *Transactions* 90 (1927): 1094–1136.

10. "The Eye-Bar Cable Suspension Bridge at Florianopolis, Brazil," *Transactions* 92 (1928): 266–342.

The reflection between deflection and vibration. Importance of the latter from the psychological as well as the structural standpoint.
Discussion of Paper #1662 by D. B. Steinman and William G. Grove. *Transactions* 92 (1928): 266–342.

11. "Studies of Shear in Reinforced Concrete Beams." *Transactions* 94 (1930): 772–73.

Lines of tension, lines of compression, and the location of cracks relative to them. Discussion of paper #1743 by T. D. Mylrea. *Transactions* 94 (1930): 734–58.

12. "Analysis of Continuous Frames by Distributing Fixed End Moments." *Transactions* 96 (1932): 1–10.

Explanation of general procedure and the mechanics of "Moment Distribution." Problem illustrating how this is an analysis by successive approximations.

Discussions
C. P. Vetter. *Transactions* 96 (1932): 11.
L. E Grinter. *Transactions* 96 (1932): 11–20.
S. S. Gorman. *Transactions* 96 (1932): 20–21.
A. A. Eremin, *Transactions* 96 (1932): 21–22.
Elmer F. Bruhn. *Transactions* 96 (1932): 22–23.
A. H. Finlay. *Transactions* 96 (1932): 23–27.
R. F. Lyman. *Transactions* 96 (1932): 27.
R. A. Caughey. *Transactions* 96 (1932): 28–29.
Orrin H. Pilkey. *Transactions* 96 (1932): 29–32.
I. Oesterblom. *Transactions* 96 (1932): 32–35.
Robert A. Black. *Transactions* 96 (1932): 35–39.
H. E. Wessman. *Transactions* 96 (1932): 40–52.
Edward J. Bendarski. *Transactions* 96 (1932): 52–55.
S. N. Mitra. *Transactions* 96 (1932): 55.
Jens Egede Nielsen. *Transactions* 96 (1932): 55–56.
F. E. Richart. *Transactions* 96 (1932): 56–60.
William A. Oliver. *Transactions* 96 (1932): 60.
R. R. Martel. *Transactions* 96 (1932): 60–66.
Clyde T. Morris. *Transactions* 96 (1932): 66–69.
Francis P. Witmer. *Transactions* 96 (1932): 69–74.
T. F. Hickerson. *Transactions* 96 (1932): 74–79.
F. H. Constant. *Transactions* 96 (1932): 79–80.
W. N. Downey. *Transactions* 96 (1932): 80–93.
R. C. Hartman. *Transactions* 96 (1932): 93–94.
Thomas C. Shedd. *Transactions* 96 (1932): 94–95.
David M. Wilson. *Transactions* 96 (1932): 96–101.
Marshall G. Findley. *Transactions* 96 (1932): 101.
George E. Large. *Transactions* 96 (1932): 101–8.
Sephus Thompson & R. H. Cutler. *Transactions* 96 (1932): 108–10.
Alfred Gordon. *Transactions* 96 (1932): 111–12.
A. W. Earl. *Transactions* 96 (1932): 112–16.
A. Floris. *Transactions* 96 (1932): 116.
I. M. Nelidov. *Transactions* 96 (1932): 117–22.

Appendix B

E. A. MacLean. *Transactions* 96 (1932): 123–24.
George M. Dillingham. *Transactions* 96 (1932): 125–27.
Donald E. Larson. *Transactions* 96 (1932): 127–33.
J. A. Van Den Brock. *Transactions* 96 (1932): 133–38.

Closure by Hardy Cross
Extension to include sidesway, when necessary. Simplification by writing only the moments carried over and making one distribution at the end. Also several other extensions and shortcuts.
Transactions 96 (1932): 138–56.

13. "The Relation of Analysis to Structural Design" Pager #1951. *Transactions* 101 (1936): 1363–74

Classification and discussion of "Deformation Stresses," Participation Stresses, Normal Structural Action and Hybrid Structural Action. Benefits of predesign.

Discussions
L. J. Mensch. *Transactions* 101 (1936): 1375–80.
Russell C. Brinker. *Transactions* 101 (1936): 1380.
Marshall G. Findley. *Transactions* 101 (1936): 1380–88.
I. K. Silverman. *Transactions* 101 (1936): 1388–89.
Bruce G. Johnson. *Transactions* 101 (1936): 1389–91.
Harold E. Wesman. *Transactions* 101 (1936): 1391–94.
N. M. Newmark. *Transactions* 101 (1936): 1325–1400.
L. E. Grinter. *Transactions* 101 (1936): 1400–1405.
F. P. Shearwood. *Transactions* 101 (1936): 1405–6.

Closure by Hardy Cross
Personal attitude toward design, which is main concept of paper.
Transactions 101 (1936): 1406–8.

14. "Interaction Between Rib & Superstructure in Concrete Arch Bridges." *Transactions* 103 (1938): 86–87.

When is analysis including affect of deck structure really worthwhile? Does the analysis indicate necessary corrections to improve design?
Discussion of paper #1981 by Nathan M. Newmark. *Transactions* 103 (1938): 62–80

15. "Preliminary Design of Suspension Bridges." *Transactions* 104 (1939): 617–18.

Function of the stiffening truss and its relation to the cable in light and heavy bridges
Discussion of Paper #2029 by Shortridge Hardesty & Harold E. Wessman. *Transactions* 104 (1939): 579–606.

16. "Rigidity & Aerodynamic Stability of Suspension Bridges." *Transactions* 110 (1945): 511–12.

Dependability and sources of evidence used in Structural Design Discussion of factors tending to "Kill" autoresonance.

Discussion of Paper #2245 by D. B. Steinman. *Transactions* 110 (1945): 439–75

Books

1. *River Control and The Yellow River of China.* Providence, R.I.: J. C. Hall, 1918.

Exhaustive compilation of published data and opinions of authorities on river control in general, and the Grand Canal and Yellow River of China in particular. Special Attention paid to movement of sediment.

2. *Statically Indeterminate Structures.* Champaign, Ill.: College Publishing, 1926.

Analysis of indeterminate structures. Introduction of the Column Analogy and Moment Distribution methods. Discussion of geometry statics, virtual work, secondary stresses and certain design factors.

3. Continuous Frames of Reinforced Concrete. New York, N.Y.: Wiley, 1932

This is not a study of all methods of analysis of indeterminate structures: Rather a discussion of the Column Analogy, Moment Distribution, geometry and statics of deflected structures, influence lines, maximum moments, shears and reactions.

4. Chapters in *Kidder-Parker Architects' & Builders Handbook.*

Chapter X, "Properties of Structural Shapes." Moment of inertia, moment of resistance, section modulus and radius of gyration.

Chapter XIV, "Strength of Columns, Posts and Struts." Timber, steel, cast iron, composite columns.

Chapter XV, "Strength of Steel Beams and Beam Griders." Bending, shearing stresses, deflection.

Chapter XIX, "Riveted Steel Plate and Box Girders." Discussion of web flanges, standards.

In the Engineering News Record

1. "Proper Form of Unsymmetrical Arch." 94, no. 17 (Apr. 1925): 702.

Letter to the Editor points out that the true line of pressure is that one lying nearest the arch axis and hence the equilibrium polygon for loads will approximate the shape of the arch axis.

2. "Temperature Deformations in Arches." 96, no. 5 (Feb. 1926): 190–91.

Method of determining temperature affects by simple calculations, which calculations are dependent upon measurement of the crown movement.

3. "Kinetic Head on Dams." 98, no. 12 (Mar. 1927): 500.

Letter to the Editor discusses the error made by many in texts on design of ma-

Appendix B

sonry dams, of increasing the head on dams because of velocity of approach. Texts in general, disagree on amount of increase. Actually it is a very small effect.

4. "Hackensack Bascule Failure" (Discussion). 103, no. 20 (Nov. 1929): 734.

Letter to the Editor discusses erroneous values obtained by the investigating board, for the forces produced by deceleration of the counter-weight on the Hackensack Bridge.

5. "Limitations and Applications of Structural Analysis." (Oct. 1933): 535–37 and (Oct. 1935): 571–73.

Objectives of analysis (research and design) and the desirable qualities for it. Use of statics and geometry. Necessity of interpretation, and consideration of element of chance.

6. "Structural Knowledge." (Feb. 1936): 199–200.

Position of structural engineering and its accomplishments throughout 1935. Review of the construction of the Transbay Bridge in San Francisco Bay. "The structural type of the year is definitely the rigid frame." Cross recognizes that the depression period has been one of originality in architectural format in structural type.

Miscellaneous Publications

1. "Temperature Deformations in Concrete Arches." *Buildings* (October 1925).

Brief method of determining strains in concrete arch due to temperature changes. Relatively inexpensive and simple.

2. Hardy Cross, "Thoughts on the Education of the Civil Engineer." *Technograph*, (University of Illinois), January 1926.

Need of discrimination in search for facts and information. Importance of questions, "hunches." Responsibility of college in training.

3. "The Civil Engineer." *Professional Engineer* (October 1928).

C.E.'s place in the world, the all-inclusiveness of his work and his development with world history. Civil engineering is not merely mathematical science; but also a field of cultural study, economic and psychological investigation and other relative values.

4. Hardy Cross, "Bridges Here and There." *Technograph* 46, no. 3, December 1931.

European and American bridges; their beauty, character and charm.

5. "Rigid Frame Bridges." In "Report of Committee on Masonry - Hardy Cross, Chairman." *Proceedings* of theAmerican Railway Engineering Association 34 (1933): 550–88.

Characteristics, uncertainties as to structural action. Procedure of general design, and analysis (particularly for LL).

Under Report of "Committee on Masonry" Hardy Cross—Chairman

6. "The Relation of Structural Mechanics to Structural Engineering." *Proceedings of the Fifth International Congress of Applied Mechanics* (1938).

Approximate methods of analysis are essential in design, to get scale on dimensions and effect of changes. However, the theory of structures is always dependent on the principals of formal mechanics; the two are, and must always be co-related.

7. "Saving Steel by Substituting Concrete." *Civil Engineering* 12, no. 7 (Jul. 1942).

Relative to war, emergency steel saving can be effected through use of other materials (timber, brick) especially concrete. Use of rigid frames, domes and shells all induce saving.

8. Hardy Cross, "Technology and the Education of Engineers." *The Yale Scientific Magazine* 20, no. 7 (April 1946).

Education of Engineers should not be considered as divorced from any other college "Department." It is related closely to science and culture and art and can't be standardized by rules, formulas, grades.

Appendix C:
Hardy Cross's Contribution to Structural Analysis

EMORY KEMP

In order to evaluate the contribution of Hardy Cross to the history of structural analysis, an insight into earlier developments is essential. The history of methods to determine moments and shears has a long and important history related to developments in structural engineering.

A significant development was the introduction of wrought iron and, later, steel, which were essentially new structural materials in the hands of architects and builders beginning with the seminal work of Henry Cort (1740–1800) and Peter Onions (fl.1780) with puddling of cast iron to produce wrought iron, and the equally important introduction of the technique of rolling structural shapes such as plates, bars, and angles. This new technology gave rise to scientific inquiries into the strength of materials and the analysis of structural forms. In a larger sense, it established a new profession of structural engineering as distinct from architecture concerned, at the time, with traditional building materials. With the advent of iron trusses, there was a serious need to determine the shapes and associated stresses in iron frame works, together with a means of determining deflections for both dead and live loads. The old empirical approach to design was no longer valid.

By the end of the 19th century another new material, reinforced concrete, entered the scene and saw rapid application in a wide variety of monolithic frameworks and slab construction. Empirical or simplified methods, although employed, were not really satisfactory and engineers called for more sophisticated analytical methods capable of determining moments and shears at any location and for a variety of loading conditions.

Several significant works have been published tracing the history of structural development from the earliest times when emphasis on the transition from an empirically based building tradition to one firmly based on scientific principles.[1] The question, in the case of trying to provide a suitable context for the work of Hardy Cross, is where to start and what developments in structural theory to leave out. Such an approach has a twofold goal, namely to identify the developments in analyses which form the background for Hardy Cross, while at the same time providing the basis for evaluating his contribution to the theoretical side of structural engineering.

Perhaps the appropriate person to start with is Galileo Gallilei (1564–1642) who provided the first attempt to ascertain the stress in a cantilever. Although his solution was not correct, it was in fact the beginning of research on the strength of materials which has had such a profound influence on those wrestling with the problem of a scientific understanding of building structures. In fact, the so-called Galileo Problem, that is the analysis of the stresses in the cantilever, was not solved correctly until 1855. This correct solution ushered in, together with other developments, the modern period of structural analysis alongside significant and rapid developments in the theory of elasticity and the strength of materials. William J. M. Rankine (1820–72) published his *Manual of Applied Mechanics in 1858*.[2] Its popularity was immense requiring nineteen subsequent editions, the last appearing in 1919 during the early career of Hardy Cross. Another Briton, James Clerk Maxwell (1830–79) published in 1864 the first analysis of indeterminate structures based on earlier work of Clapeyron (1799–1864) and later improved by Otto Mohr (1835–1918). This method is now known as the Maxwell-Mohr method, since Mohr presented a simplified, and at the same time more general, derivation of Maxwell's work. Shortly after Maxwell's publication, Alberto Castigliano (1847–84) published in 1873 his theorems on strain energy also know as the method of least work. His paper appeared in 1858.[3]

What was clearly needed was a method or methods to solve highly indeterminate structures such as engineers were increasingly encountering in reinforced concrete frameworks and in tall buildings and long span-bridges. A structure is said to be indeterminate when there are more unknown stress resultants (stress resultants include moments, shears, torsion, for example) than can be determined by the basic equations of equilibrium.

In 1852, Mohr published a procedure for determining the secondary stresses in trusses which laid the foundation for the slope-deflection method. This was followed in 1914 with a paper written by Axel Bendixen[4] which can be considered to be the basis of the slope-deflection method using the well known equations for moments and rotations in any member m-n, namely;

$M_{mn} = 2k_{mn} (2\ominus_{mn} + \odot_{nm}) = M°_{mn}$
M_{mn} = the bending moment acting on end m of bar mn
$\ominus_{mn}$ and $\odot^{nm}$ the angles of rotation of the ends of beam mn
$k_{mn} = EI_{mn}/\ell_{mn}$ and is called the stiffness factor where I is the moment of inertia, E is the modulus of elasticity of the beam, while ℓ_{mn} is the length of the beam.

In the following year, 1915, George A. Maney of the University of Minnesota published an independent version of the slope-deflection method, which was widely employed by structural engineers until it was eclipsed by Hardy Cross's moment distribution method, made available in the 1930s.[5]

The slope-deflection method utilizes joint rotation and translation and as such, can address the problem of deflections of a support due to settlement or other movements making the method quite versatile without adding any complexities.

Perhaps the best means to explain the method is to give an example. The two span beam shown in Figure 18 has a constant cross-section and is loaded uniformly in one span only. The exercise is to determine the resulting moment over the interior support "B." The basic slope-deflection equations are applied to positions "A" and "C" at the ends of the continuous beam. In addition, a pair of equations is written for the moment at either side of the middle support and noting that these moments must balance each other.

The method works efficiently with linear members of constant cross-section. The solution for non-uniform cross-section members is unusually complicated and has little appeal for practicing engineers.

Into this stimulating field in structural analysis, Hardy Cross began his academic career. Apparently as early as 1924, Cross was teaching his new method, appropriately called moment distribution, but it was not published until 1930. The method is one of the iterations of moments at successive joints rather than solving a series of simultaneous equations for the unknown stress resultants. The iterated method had a decided advantage for structural engineers in the days before computers. The iteration is achieved by imagining all of the joints in a continuous beam or framework as locked and subjected to appropriate fixed end moments. Each of the joints is released successively. The moments balanced and the released joint and the carry-over moments distributed to adjacent joints. The procedure is repeated as successive joints are released, moments balanced, and carry-over factors distribute the moments to the opposite ends of each member. Before moment distribution can be undertaken, the fixed end moments need to be determined, and for the case of non-uniform members, the carry-over factor as well as the fixed end moments must be determined. For members with a constant moment of inertia and modulus of elasticity, the carry-over factor is one-half. This can be easily demonstrated by the slope deflection or other methods. For non-prismatic members, both the fixed end moment and carry-over factor must be calculated using other methods such as Hardy Cross's column analogy. The simultaneous equations associated with the moment distribution method are typically well conditioned and the iteration procedure has rapid convergence.

By considering the same example as used for the slope-deflection method, the essential features of the moment distribution method can be seen and details compared to the earlier slope deflection solution (see Figure 18, the method is further illustration in Figure 19).

Before leaving the moment distribution method, it is important to note that Professor Richard C. Southwell, of Oxford University in England, also developed an iterative procedure for analyzing structural frameworks. The method, called the Relaxation Method, most notably served as the basis for the design of the three-dimensional metal framework of British R101 dirigible. Published in 1935, Southwell's

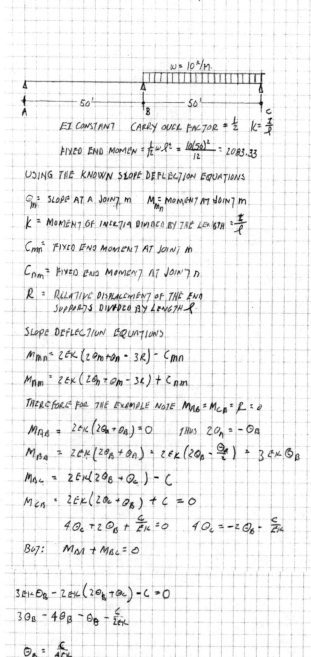

Figure 18. Example of the analysis of a two span beam by the slope deflection method. (E. L. Kemp)

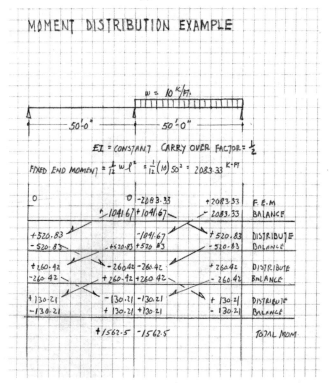

Figure 19. Example of the analysis of a two span beam by the moment distribution method. (E. L. Kemp)

initial paper on the subject was followed by his well-known book entitled *Relaxation Methods in Engineering Science* and published in 1940.[6] Southwell discusses the comparison between the moment distribution method and the more general Relaxations Method for continuous beams with rigid supports and demonstrates that the methods in this case are identical. The Cross method must be modified when sidesway results in a bent or framework and does not directly deal with joint movements, say the settlement of a support. For this case, the slope-deflection method provides the most straightforward solution. Although more general and theoretical, the Southwell method never achieved the notoriety of Cross's moment distribution method.

Legend has it that Hardy Cross, the master teacher, discovered the column analogy method when a student was writing the following equations on the blackboard:[7]

Angle between tangents,

$$\phi = \int m_i \frac{ds}{EI}$$

Vertical deflection of one end from the tangent at the other end

$$\Delta_y = \int m_i x \frac{ds}{EI}$$

Horizontal deflection of one end from the tangent at the other end

$$\Delta_x = \int m_i y \frac{ds}{EI}$$

m is the indeterminate moment at any point caused by restraint. The "static" moments caused by the external loading on the beam when the latter is in a

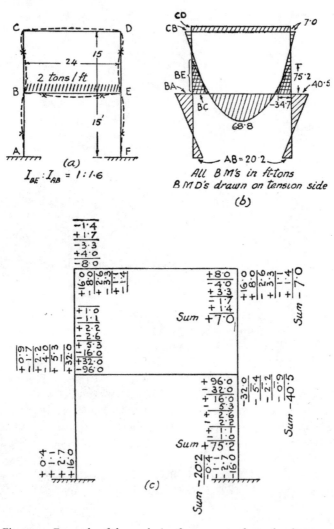

Figure 20. Example of the analysis of a two story frame by the moment distribution method. (Credit: W. Fisher Cassie, *Structural Analysis*, 143).[8]

Appendix C

statically determinate condition (resulting from the release of restraints) must be accounted for separately. The rotation of a fixed end is zero, and the φ must be counteracted by an opposite rotation caused by the loading on the beam when it is in the statically determinate condition. This rotation

$$= \int m_s \frac{ds}{EI}$$ where m is "static" moment at any section.

Cross recognized that they are exactly the same form as the well-known;

Equations for a Short Column eccentrically loaded
Total load on the column, $P = \int p.dA$.
Moment about the Y axis, $M_y = \int px.dA$

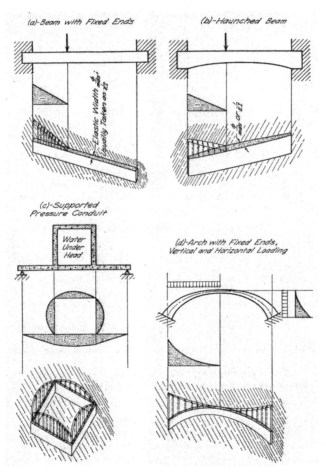

Figure 21. Types of analogous column sections. (Credit: Hardy Cross, *Column Analogy*, 26)

108   Appendix C

Moment about the X axis, $M_z f = py.dA$.
p is the fibre stress at any point of the column section.

If the indeterminate structure is released, then the equivalent column analogy load is the area of the moment diagram divided by EI. The area of the moment diagram is the load on the head of the analogous column acting through the centroid of the moment over EI diagram multiplied by the eccentricity of the equivalent short

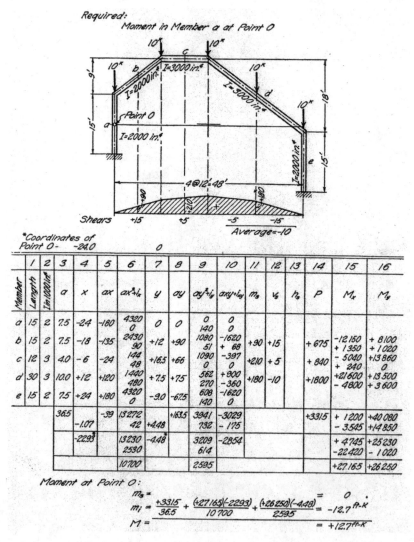

Figure 22. Unsymmetrical bent analyzed by the column analogy method. (Credit: Hardy Cross, *Column Analogy*, 40)

analogous column. Since engineers were and are familiar with eccentrically loaded columns, the analogy provided a powerful and yet familiar procedure for analyzing single ring indeterminate structures. By "single ring" we mean those structures which are part of a ring such as variable cross-section beams, arches, and frameworks. Included in his bulletin on column analogy, Cross presents several examples including an unsymmetrical bent. Thus, multi-span beams and many frames cannot be solved directly by this method.

For many engineers, perhaps the most useful feature of the column analogy method is the determination of fixed end moments and carry-over factors for non-prismatic members, which can then be used with the moment distribution system. The two methods can thus be closely linked for the solution of more complicated structures. The method is also well adapted for haunched beams and rigid frame bents, which were popular in the 1930s and 1940s for highway bridges. The following examples demonstrate how the method can be used to calculate fixed end moments for a beam loaded uniformly and the case of the single-concentrated load. Regardless of the shape of the beam, the carry-over factor can be determined by placing a

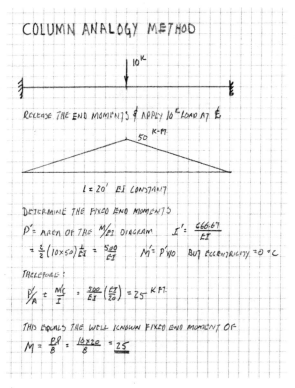

Figure 23. Calculation of the fixed-end moment for a beam with a central point load. (E. L. Kemp)

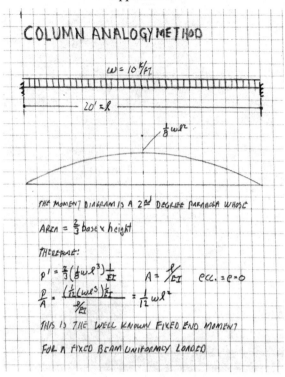

Figure 24. Calculation of the fixed-end moment for a uniformly distributed load. (E. L. Kemp)

one radian load at the end and determining the transfer of moment to the other end of a given member.

By the 1930s, the Cross moment distribution coupled with the column analogy method had become the most popular analytical tools used by structural engineers.

Notes

Preface

1. Theodore von Karman and Lee Edson, *The Wind and Beyond* (Boston and Toronto, 1967), 55.
2. Ibid., 127.
3. Kingsbury, Berg, and Schillinger, *Men and Ideas in Engineering,* (Urbana, 1966), 10.

Chapter 1: The Preparatory Years, 1885–1921

1. This description of Nansemond County is taken from James B. Dunn, *The History of Nansemond County* (Suffolk, Va., 1907). Material on the Cross family comes from the brochure of an exhibit at the Isle of Wight County Museum in Smithfield, Va., in 1992. The creator of this exhibit, Ms. Dinah Everett, has been most helpful in checking the deed book of Nansemond County for the transactions mentioned herein. Professor Richard Guy Wilson of the University of Virginia ascertained the attendance of Thomas Hardy Cross at that University in 1858–59.
2. Dunn, *History,* 7. Colonel John W. Thomason, the noted writer and illustrator, set a chapter on Nansemond County in *Lone Star Preacher* (New York, 1941). Thomason related that General Lee, in order to feed his men, dispatched the Texas Brigade and Pickett's division to forage in the Virginia and North Carolina tidewater counties in the spring of 1863. Agents had reported a fine crop year in the region below Suffolk. Quantities of corn and bacon were available. The story revolves around the ambush of a contingent of federal cavalry by the local home guard.
3. Thomas J. Wertenbaker has dealt with the history of Norfolk Academy in *Norfolk, Historic Southern Port* (Duke University Press, Durham, 1962). The academy itself has been most generous in supplying me with documents. Particularly valuable were two pamphlets: *Norfolk Academy: A Private School for Boys and Young Men* (July, 1898) and *Semi Annual Report of the Work Done by Students of Norfolk Academy for the First Half of the School Year 1898–99.*

4. *Norfolk Academy*, pamphlet, 4.

5. *Norfolk Academy Semi Annual Report 1898–99*, 3.

6. In the autumn of 1957 Cross was interviewed by two students from the *Belfry*, a student publication of Norfolk Academy. These excerpts are from that interview.

7. Letter from John Brinkley, College Historian of Hampden-Sydney, 18 November 2000. Cross was always loyal to his old college. At his death one-third of his estate went to Hampden-Sydney.

8. Cross's transcript at M.I.T. was furnished to me in a letter of 2 November 2000.

9. P. Reyner Banham, *A Concrete Atlantis*, (Cambridge, 1986), 171–77.

10. Cross's transcript at Harvard University was furnished to me in a letter of 18 October 2000.

11. Brown University supplied reproductions of its catalogue offerings in Civil Engineering for the years 1911–18. These course descriptions are taken from those catalogues.

12. On September 5, 1921, Cross married Edythe Hopwood Fenner of Providence, R.I. From all accounts she was charming but frail. They had no children. Robert Goodpasture, who knew the Crosses in New Haven, describes them as "a devoted couple."

13. This brief characterization is by William F. Uhl in *Charles T. Main: One of America's Best* (Newcomen Society, Boston, 1951), p. 17. This brief pamphlet is a Newcomen Society address. Uhl knew Main well and at the time was president of Charles T. Main Inc.

Chapter 2: The Creative Years at Illinois, 1921–37

1. Jack C. McCormac, *Structural Analysis*, (Scranton, 1962), 310.

2. This anecdote, and the following pages on Cross's colleagues, are based on Kingsbury, Berg and Schillinger, *Men and Ideas in Engineering*.

3. Hardy Cross, *Notes on Statically Indeterminate Structures*," (Champaign, Ill.), 1926, 1.

4. Ibid., 2.

5. Ibid.

6. Ibid.

7. Ibid.

8. David Billington, *Robert Maillart's Bridges: The Art of Engineering* (Princeton, 1979), 141.

9. Thomas Golden to Leonard Eaton. 4 December 2001.

10. Cross, *Notes*, 3.

11. Ibid.

12. Ibid.

13. *Arches, Continuous Frames, Columns, and Conduits: Selected Papers of Hardy Cross*. Introduction by Nathan M. Newmark (Champaign, 1963), 233. Cross generally disliked footnotes, and perhaps regrettably, used none in *Engineers and Ivory Towers*.

14. Ibid., 240.

15. Ibid., 164.

16. Ibid.

17. P. Reyner Banham, *A Concrete Atlantis: U.S. Industrial Building and Modern Architecture* (Cambridge, Mass., 1989). See especially the introduction and 23–107, "The Daylight Factory."

18. Carl Condit, *American Building* (Chicago, 1968), 243.

19. Cross, *Arches*, 110.

20. H. M. Westergaard and W. M. Slater, "Moments and Stresses in Slabs," American Concrete Institute, *Proceedings, Vol. XVII* (1921). Reprinted by the National Science Council, *Report*, Series 32. I have devoted some attention to Maillart as a slab designer in "Frank Lloyd Wright and the Concrete Slab and Column," The *Journal of Architecture*, vol. III, Winter, 1998, 315–46.

21. Cross, *Arches*, 2.

22. Ibid., 2.

23. *Impressions of Mies: An Interview of Mies Van Der Rohe, His Early Chicago Years*, ed. William S. Shell (Knoxville, 1988), 17.

24. Stefan Timoshenko, *History of Strength of Materials* (N.Y., 1953), 424.

25. McCormack, *Structural Analysis*, 329.

26. I owe this discovery to Ms. Brooksie Koopman of Alexandria, Virginia.

27. Meredith Clausen, "Belluschi and the Equitable Building in History," *Journal of the Society of Architectural Historians*, (1992), 109–29.

28. Nathan Newmark in the introduction to *Arches*, X.

29. Norbert Wiener, *I Am a Mathematician* (Cambridge, 1956), 258.

Chapter 3: The Years at Yale, 1937–51, and Retirement

1. For Terzaghi's version of the episode, see Richard Goodman *Karl Terzaghi, The Engineer as Artist*. Washington 1999, 188–232.

2. John Salvaggio, M. D., *New Orleans Charity Hospital: A Story of Physicians, Politics, and Poverty* (Louisiana State University Press, Baton Rouge, 1992), 142.

3. Ibid., 143.

4. Theodore von Karman and Lee Edson, *The Wind and Beyond* (Boston, 1967), 214.

5. Mario Salvadori and Matthys Levy, *Why Buildings Fall Down*, (New York and London, 1992), 118–19.

6. Henry Petroski, *Design Paradigms: Case Histories of Error and Judgment in Engineering* (Cambridge, 1994), 144–63.

7. David B. Steinman and Sara Ruth Watson, *Bridges and their Builders* (New York, 1941), 385–86.

8. H. W. Coulter, *Theory of Structures*, (London, 1959).

9. Hardy Cross, *Engineers and Ivory Towers* ed. Robert C. Goodpasture (New York, 1951), 21.

10. Ibid., 21–22.

11. Ibid., 24.

12. Ibid., 26.

13. Ibid., 32.

14. Ibid., 35.

15. Ibid., 38.

16. Bernard Morrey, *The Dictionary of Art Online.www.Groveart.com*.

17. Letter to the writer from Robert C. Goodpasture.

18. Kuesel, *Notes on Cross*, I, 31.

19. Ibid., 2.

20. Ibid., 37. Mr. Kuesel characterizes this remark as "Pure Cross." Telephone conversation, 26 February 2002.

21. Ibid, 28.

22. Ibid., 1.

23. Ibid, 20.

24. Ibid., 37.

25. Kuesel, *Notes,* II, 25.

26. Ibid., 39.

27. Ibid., 42.

28. Ibid., 98.

29. Ibid., 115.

30. Hardy Cross, *Engineers,* p. 41.

31. Telephone conversation with Professor Emeritus Charles Sawyer, September 2001.

32. W. Jack Cunningham, *Engineering at Yale* (New Haven, 1992), 33.

33. Ibid., 44.

34. Ibid.

Appendix C

1. Stephen P. Timoshenko, *History of Strength of Materials,* (New York, 1983), 304–27.
J. Stearling Kinney, *Indeterminate Structural Analysis,* (Reading, Mass., 1957), 1–19.
Hans Straub, *A History of Civil Engineering,* English translation, (Cambridge, Mass., 1964), 173–86 and 222–23.

2. Timoshenko, *Strength of Materials,* 197–98.

3. Ibid., 289.

4. Ibid., 422.

5. Hardy Cross, "Analysis of Continuous Frames by Distributing Fixed End Moment," *Pro. Am. Soc. Civil Engrs.,* May 1930.

6. Richard V. Southwell, *Relaxation Methods in Engineering Science,* (London, 1951), 71–99.

7. Hardy Cross, *The Column Analogy,* Bulletin 215 (Urbana, Ill., University of Illinois Engr. Exp. Sta., 1930).

8. W. Fisher Cassie, *Structural Analysis* (London, 1954), 175.

Index

Note: Illustrations are indicated by boldface type.

Aalto, Alvar, 25
Adler & Sullivan (architects), 12
aerodynamics, 62
aeronautical engineering, 61
aircraft design, 41
air distribution, 44
Alford (student of Hardy Cross), 20–21
American Concrete Institute, 34, 53
American Railway Engineering Association, 46–47, 54
American Society of Civil Engineers (ASCE), 30, 39, 53
Amman, Othar, 61
analysis, 29–30. *See also* continuous structure analysis; Cross method
"Analysis of Continuous Frames by Distributing Fixed-End Moments" (Cross), 39, 43, 62
Analysis of Engineering Structures (Pippard, Baker), 62–63
"Analysis of Flow in Networks of Conduits or Conductors" (Cross), 44–45, 85
Archangliska, Alexandra, 42
Arches, Continuous Frames, Columns, and Conduits (Cross, Newmark), 23, 31, 35–36
arches (diagram), **80**
A.S.C.E. (American Society of Civil Engineers), 30, 39, 53

Bagby, J. H. C., 9–10
Baker, John Fleetwood, 62–63
Banham, Reyner, 13, 32
Barr, Alfred, 33
Barren Ground (Glasgow), 2
Bauhaus, 33
Der Bauingenieur (Dernedde), 64
beam theory, 25, 38
Belluschi, Pietro, 48–50
Bendixen, Axel, 19, 23, 37
Berlage, J. P., 63
"Big Four," 21–23
binomial theorem, 44
Boulder (Hoover) Dam, 22–23
The Bridge Engineer as a Mathematician (Steinman), 61–62
"Bridges Here and There" (Cross), 68–70
Brown University School of Engineering, 13–16
Bulletins (University of Illinois Experiment Station), 31–32, 44

"Calculation of Continuous Frames by the Method of Moment Distribution" (Fornerod), 64
California Institute of Technology (Cal Tech), 61
Calisev, K. A., 37

Index

Cal Tech (California Institute of Technology), 61
Captain James B. Eads bridge, 12
Charity Hospital (New Orleans, Louisiana), 54–60; conclusions reached in study of, 59; corruption in construction of, 57–58; Cross report concerning, 54–57
Clapeyron, B. E., 18
Clausen, Meredith, 48–49
Column Analogy, 24, 29
The Column Analogy (Cross), 31–32
column and slab concrete, 33–34
Conant, James Bryant, 82
concrete. *See* reinforced concrete
Condit, Carl, 32
conservatism, 66
continuity, 18, 25
Continuous Building Frames of Reinforced Concrete (Cross, Morgan), 23, 83
continuous frames, 34, 35–36, **36,** 37
continuous structure analysis, 47–48
Cross, Edythe Fenner (wife), 83
Cross, Eleanor Elizabeth Wright (mother), 4, 10
Cross, Hardy, **8, 20, 86;** Americanism of, 52; awards won by, 53, 82, 86–87; birth and childhood of, 4; Boston engineering work of, 16–17; as bridge engineering connoisseur, 37; Brown University teaching of, 13–16; college undergraduate education of, 7–9; conservatism of, 66; consultation work of, 54–61; as devil's advocate, 20–21; drawing pictures and, 28; earthquake resistant design and, 51–52; elasticity theory contributions of, 15–16; elementary education of, 4; as engineer-philosopher, 50–51, 52; environmental influences upon, 1–4; European travels of, 30; foresight of, 16; Harvard University degree of, 13; hobbies of, 86; honorary degrees awarded to, 53; as humanist, 83; hydraulic engineering contributions, 44–45; impact of, 48–50, 51; intellectual qualities of, 7, 23, 50–51; international recognition of, 62–65; knowledge about engineering developments nationwide, 45–46; liberal education of, 66–67; materials treated by, 32; methodological preferences of, 16; Missouri Pacific Bridge Department work experience, 12–13; M.I.T. education, 11–12; Norfolk Academy teaching of, 11, 13; practicality of, 37; pragmatism of, 52; principles of statical indeterminacy developed by, 29–30; professional committee memberships of, 53–54; professional objectives summarized, 7; retirement years of, 85–87; secondary education of, 4–7; skepticism of, 85; teaching methods of, 71–78; temperament of, 54, 59, 83–84; textbook based on notes from lectures of, 23–24; as utilizing simplest means, 28–29, 40; work ethic of, 23; writing style of, 23–24, 30, 38; Yale University tenure of, 53–54, 67, 71–73, 79, 83–85. *See also* Cross method; Cross quoted
Cross, Thomas Hardy (father), 1–3, 10
Cross, Thomas Peete (brother), 4, 10–11, 85
Cross method: approximation in, 39; beam constants in, 38; buildings designed using, 87; bypass of adjusting rotations in, 38; comparative ease of, 38–39; computer programs and, 44; diagrams of, 40, 41; first presentation of, 24; as gathering of ideas, 39–40; importance of, 39, 43–44; locked structures concept and, 44; mid-century history of, 48–50; practicality of, 39; questions about, 42. *See also* moment distribution
Cross quoted: on bridges, 68–70, 81–82; on civil engineering profession, 78; on current structural knowledge developments, 45–48; on development of field of engineering, 81; on education, 7; on elasticity theory, 26, 27; on geometry, 27–28, 32, 44; on medieval architecture, 65; on nature of theory in engineering, 26; on Norfolk Academy, 7; on philology, 85; on questions, 75; on skepticism and doctrine, 76; on standardization, 65–66; on structural design limitations, 28–29; on teachers, 67–68; on teaching engineering, 79; on term definition, 76; on wisdom in engineering, 76. *See also* Cross, Hardy
Das Cross-Verfahren (Johannson), 64
Culmann, Karl, 28
Cunningham, W. Jack, 84–85

Davenport Bridge, 46, 47–48
Dawson Warehouse, 25

Index

Dernedde, W., 64
Der Rohe, Mies van, 33, 41, 42
Dialogues Concerning Two Sciences (Galileo), 26
distribution of flow in sample network (diagram), **45**
Distribution of Moments, 29–30
Dudley, Samuel W., 84

earthquake impact on structures, 22
earthquake resistant design, 46, 51–52
Eddington, Arthur, 50
education, 4–7, 66–68. *See also* Cross quoted
Eiffel, Gustav, 51
Einstein, Albert, 39–40, 52
elasticity theory, 15–16, 26–27
electrical circuits, 44
Emerson, Ralph Waldo, 52
Engineering: A Weekly Journal, 62
Engineering News Record, 37
engineer-philosopher, 50–51, 52
Engineers and Ivory Towers (Cross), 66, 71–72
Equitable Building (Portland, Oregon), 48–50
"Experimental Studies of the Lateral Thrust and Angle of Internal Friction of Soils" (Cross), 11–12

failure analysis, 78
"Farmer's Delight," 1–2
Farquaharson, F. B., 60
The Ferro Concrete Style (Onderdonk), 25
fiber stress theory, 25
flat slab, 82
Fornerod, M. F., 63–64
Franciscan Hotel, 25
Franklin Institute, 87
Freyssinet, Eugene, 62, 71, 74
Fujikawa, Joe, 41

Gaczat, Gunter, 64–65
Galerie des Machines, 34
Galileo, 26
"Galloping Gertie," 60–61
geometry, 27–28, 32, 44, 50
George A. Fuller & Company, 58
Giedion, Siegfried, 25, 62
Glasgow, Ellen, 2
Golden, Thomas, 1, 28

Golden Gate Bridge, 46, 51, 60
Gold Medal of Institution of Civil Engineers of Great Britain, 82, 86–87
Goodpasture, Robert, 28, 71, 83
graphic statics, 28
Greene, Lockwood and Co., 32
Grinter, Lawrence P., 42, 65
Griswold, A. Whitney, 84–85
Gropius, Adolf, 33

Hall, William J., 24
Hampden-Sydney College, 7–9, 82
Hardy Cross method, 44–45
Harvard University, 13, 82
Heisenberg, Werner, 52
Hennebique, François, 12
High Bridge at Farmville, Virginia, 9, **10**
high-strength bolts, 22, 41
History of Strength of Materials (Timoshenko), 42, 43
Hitchcock, Henry-Russell, 33
Hooke, Robert, 26
Hoover (Boulder) Dam, 22–23
Hotel Palacio Salvo, 25
Hunley, J. B., 46–47
Hyatt Regency Hotel (Kansas City, Missouri), 78–79
hydraulic engineering, 44–45

improved flat slab, 82
"Improved Methods of Calculation for Continuous Beams and Frames" (Dernedde), 64
Ingalls building (Cincinnati, Ohio), 12
Institution of Civil Engineers of Great Britain, 82–83, 86–87
"Interaction of Rib and Superstructure in Concrete Arch Bridges" (Newmark), 79–80
The International Style (Hitchcock, Johnson), 33

James, William, 52
Johannson, Johannes, 64
John Hancock Center, 51
Johnson, Philip, 33

Kahn, Albert, 32
Karman, Thodore von, 61

Ketchum, Milo, 39
Khan, Fazlur, 51
Khatchaturian, Narbey, 23
Klein, Felix, 19
Kornacker, Frank, 41
Kuesel, Thomas, 61, 72–78

Lamme Medal (American Society of Engineering Education), 53
Langley Aeronautical Laboratory, 41
Lee Paterson Bridge, 62
Lehigh University, 47
Leland Stanford Museum, 12
Levy, Matthys, 61
"Limitations and Application of Structural Analysis" (Cross), 50–51
Long, Huey, 58
Lundquist, Eric, 41

Magnel, Gustave Paul Robert, 62, 74
Mahone, W. T., 2–3
Maillart, Robert, 33, 34, 37, 63, 79
Main, Charles T., 16–17
Massachusetts Institute of Technology (M.I.T.), 11–12, 47, 53
Maxwell, James Clerk, 7, 26–27, 30, 87
McCormac, Jack C., 19, 43
McCullough, Conde, 62
medieval architecture, 65
metal fatigue, 22
La Methode de Hardy Cross et Ses Simplifications (Zaytzeff), 63
Meyer, Adolf, 33
Mickal, Abe, 58
Missouri Pacific Railroad, 12–13
M.I.T. (Massachusetts Institute of Technology), 11–12, 47, 53
Mohr, Otto, 30, 87
Moissieff, Leon, 61
moment distribution, 31–32
moment distribution (diagram), **40**
moment distribution method. *See* Cross method
moment weight computation (diagram), **31**
moment weights (trusses), 30–31
Morgan, Newell D., 23, 82

Nansemond County, Virginia, 1–4

National Committee on Suspension Bridges, 53
nature of engineering theory, 26
Nervi, Pier Luigi, 79
Newmark, Nathan, 23, 29, 79
Newton, Isaac, 26, 28
Norfolk Academy, 4–7, 11, 13, 81
Norman Medal (American Society of Civil Engineers), 53
Norton, William Augustus, 53
Notes on Cross (Kuesel), 73–79
Notes on Indeterminate Structures (Cross), 23–24
Notes on Solid Geometry (Venable), 9
Numerical Methods of Analysis in Engineering (Cross, Grinter), 42

Objections to Continuous Beams (Cross), 80–81
Onderdonk, Francis, 25

Pacific Borax factory, 12
Paimio Sanitorium, 25
Pan Am building, 22
Pascal, Blaise, 40
pavement slab analysis, 22
Petersburg Technological Institute, 43
Petroski, Henry, 61
pipe networks, 44
Pippard, Alfred John Sutton, 62–63
Prak, Niels, 33
Prestressed Concrete; Its Principles and Applications (Magnel), 74
principles of statical indeterminacy analysis, 29–30
Proceedings (American Concrete Institute), 34
Proceedings (A.S.C.E. journal), 30, 39
The Provincial Letters (Pascal), 40

"railroad transition spiral," 22
Ransome, Ernest L., 12, 32
Rayleigh, Lord, 42–43
reinforced concrete, 12, 32–33, 78. *See also* continuous structure analysis; Cross method; elasticity theory; rigid-frame structure
rigid-frame structure, 37, 47–48

River Flow Phenomena and Hydrography of the Yellow River in China (Cross), 19
Roosevelt, Franklin D., 58, 61

Salginatobel Bridge (Maillart), **36,** 79
Salvadori, Mario, 61, 78
San Francisco-Oakland Bay Bridge, 46
Der Schweizerische Bauzeitung, 63, 82
Seagram Building, 41, 42
Sears Tower, 51
Seymour, Charles, 84
skyscrapers, 51. *See also* specific buildings by name
Slater, W. A., 33
slope deflection, 19, 37
soil mechanics, 59–60
"Some Recent Developments in Methods of Analyzing Statically Indeterminate Structures" (Wilson), 23
"Stability of Structural Members Under Axial Load" (Lundquist), 41
standardization, 65
Stanford University, 43, 47
statical indeterminacy, 16; Cross analysis methodology for, 29–30; Cross involvement with, 16; Cross research and breakthrough in, 23; definition of, 18; fiber stress theory and, 25; first method for analyzing, 27; fundamental principle of, 25; geometry and, 27–28; as structural engineering issue, 24; types of structures characterized by, 24–25. *See also* Cross method
steam distribution, 44
steel railway bridges, 46–47
Steinman, David, 61–62
Structural Analysis (McCormac), 43
structural design limitations, 28–29
"Structural Knowledge" (Cross), 45–48
structural mechanics, 22
Strutt, John William. *See* Rayleigh, Lord
Swain, George E., 30

Tacoma Narrows Bridge, 60–61, 62
Talbot, Arthur Newell, 21
"Talbot spiral," 22
teachers, 67–68
teaching methods, 71–78

La Technique Moderne—Construction (Zaytzeff), 63
The Technograph (University of Illinois at Urbana-Champaign College of Engineering periodical), 68–70
Terghazi, Karl, 54, 59–60
Theorem of the Three Moments, 18–19
theoretical and applied mechanics, 21, 22
theory of elasticity, 15–16, 26–27
The Theory of Elasticity (Timoshenko), 38
theory of fiber stress, 25
The Theory of Sound (Rayleigh), 42–43
"Thin-Section Concrete Arches as Built in Switzerland" (*Engineering News Record*), 37
timber bridges, 47
Times Picayune (Cross report publisher), 54–57
Timoshenko, Stefan, 42–43, 81
Torroja, José Antonio Eduardo, 79
Tredgold, Thomas, 82
Triborough Bridge, 46
Tunstall, Robert W., 4–5, 6, 7
Turner, Claude A. P., 33, 63, 82

Union Trust building, 12
United Shoe Machinery factories, 12
University of Illinois at Urbana-Champaign, 19–20, 21–22, 23
University of Illinois Experiment Station, 31–32, 44
University of Michigan, 43
University of Washington, 60
U.S. Air Force Academy, 44

Van Nelle Factory, 33, **34**
Venable, C. I., 9
Verrazano Bridge, 22, 41
Des Verwollstandigte Cross—Verfahren in der Rahmen Berechnung (Springer-Verlag), 64
Vierendeel truss, 63
"Virtual Work: A Restatement" (Cross), 30–31, 39

Wainwright Building, 12
Walnut Lane Bridge, 62
Walter, Thomas U., 4
Wason Medal (of the American Concrete Institute), 53

water distribution systems, 44–46
Weiss, Dreyfous, and Seiferth, 58
Westergaard, Harald, 12, 21, 22, 33, 85
Westinghouse Corporation, 43
"Why Continuous Frames?" (Cross), 34
Wiener, Norbert, 52
William Ward house, 12
Wilson, Wilbur, 19, 21, 22, 23, 34, 86
Wohlenberg, Walter J., 84

Wolfe, Tom, 53
Works Progress Administration (W.P.A.), 58
W.P.A. (Works Progress Administration), 58
Wright, Frank Lloyd, 63

Yale Engineering Association, 85
Yale University, 53–54

Zaytzeff, Serge, 63

LEONARD K. EATON is Emeritus Professor of Architecture at the the University of Michigan. Well known for his studies of Frank Lloyd Wright, Professor Eaton has published numerous works, including *Landscape Artist in America: The Life and Work of Jens Jensen, Two Chicago Architects and Their Clients, American Architecture Comes of Age,* and *Gateway Cities and Other Essays.*

The University of Illinois Press
is a founding member of the
Association of American University Presses.

_____

Composed in 10.5/13 Adobe Minion
by Type One, LLC
for the University of Illinois Press
Manufactured by Thomson-Shore, Inc.

University of Illinois Press
1325 South Oak Street
Champaign, IL 61820-6903
www.press.uillinois.edu